Managing Safety in the Chemical Laboratory

Managing Safety in the Chemical Laboratory

James P. Dux, Ph.D.
Robert F. Stalzer, Ph.D.

VNR VAN NOSTRAND REINHOLD COMPANY
New York

Library of Congress Catalog Card Number 87-23043

ISBN 0-442-21869-9

Printed in the United States of America

Van Nostrand Reinhold Company Inc.
115 Fifth Avenue
New York, New York 10003

Van Nostrand Reinhold Company Limited
Molly Millars Lane
Wokingham, Berkshire RG11 2PY, England

Van Nostrand Reinhold
480 La Trobe Street
Melbourne, Victoria 3000, Australia

Macmillan of Canada
Division of Canada Publishing Corporation
164 Commander Boulevard
Agincourt, Ontario MIS 3C7, Canada

16 15 14 13 12 11 10 9 8 7 6 5 4 3 2 1

Library of Congress Cataloging-in-Publication Data

Dux, James P.
Managing safety in the chemical laboratory.

Bibliography: p
1. Chemical laboratories–Safety measures. I. Stalzer, Robert F. II. Title.
QD51.D89 1988 542'.028'9 87-23043
ISBN 0-442-21869-9

Contents

Preface

"What—another book on safety?" If that was your reaction when you first saw this book, you missed the most important word in the title, the first one: *managing.* This book was written with the express purpose of benefiting persons with overall responsibility for laboratory management, or whose major responsibility is safety in laboratory operations.

There are many books available on safety. Unfortunately, most of them are written for manufacturing plant operations, and not for laboratories. The laboratory, and especially the chemistry laboratory, is a unique environment with its own specific safety considerations, problems, and solutions. For example, the manufacturing worker may be exposed to a few chemicals in large quantities. The chemical laboratory worker, on the other hand, is exposed to a large variety of chemicals, but usually only in small quantities at any given time. In addition, the chemicals used in a laboratory may often be so hazardous (with regard to toxicity or explosive potential, for example) that they could be accepted in a manufacturing operation only with many engineered safety devices.

There are, of course, books available on laboratory safety, and we shall be referring to them in this text. However, a perusal of them will reveal that they tend to focus on the "nuts and bolts" approach to safety: the equipment and mechanical procedures of safe laboratory practices. In our opinion, what is needed is a book that can offer guidance in *managing* safety in the chemical laboratory.

Chemists, and the technicians who work with them, tend to be people who are oriented toward "things"—equipment, apparatus, instruments, and the like. Their natural approach to a safety problem is a "technical fix": find a gadget that can prevent it from happening again. Rarely is thought given to the underlying situation that caused the accident in the first place, which more often than not may be a management problem, such as insufficient instruction, poor work habits, or lack of preparation for an emergency.

The authors' experience in working in chemical laboratories spans four decades. We can attest to the fact that there has been a tremendous improvement in the attitude toward laboratory safety in those years, as well as in the safety equipment available. Yet accidents still occur in the laboratory, and the time seems ripe for a close look at improving laboratory safety by application of improved management.

Improved safety management will not only benefit the laboratory worker, but also the manager and the organization of which he or she is a member. Although benefits of an improved safety record may be difficult to identify on the accountant's profit/loss statement, they are nevertheless real in terms of improved employee morale, lower insurance costs, and—most important of all—costs of "disasters averted."

There is an old saying that experience is the best teacher. That is true, but experience also gives the exam before the lesson is taught. With the help of this book we hope that laboratory managers can learn the lesson before they are put to the test.

This book has been written mainly for managers in small and medium-sized laboratories engaged in the more common chemical operations. Highly specialized technologies, such as lasers, X-ray analyses, radiochemistry, or biochemistry, have not been considered in detail. We refer the reader to specialized texts for information on safety in these types of operations.

Acknowledgments

The authors gratefully acknowledge the following individuals for their assistance and helpful suggestions in the preparation of this book:

Dr. Charles E. Garland for his comments and suggestions for chapter 5.

Dr. Thomas G. Greco for his suggestions for chapter 9.

Mr. Christopher P. McCready for his suggestions for chapter 10.

Dr. W. G. Mikell for his general review of and suggestions for the entire manuscript.

1

Management and Laboratory Safety

HISTORICAL REVIEW

Tucked away in the hills of north-central Pennsylvania, on the banks of the Susquehanna river, lies the small town of Northumberland. The major point of interest for the chemist visiting Northumberland is the former residence of Joseph Priestly, one of the great chemists of the eighteenth century. The Priestly house has been restored to the condition it was in in his day, and contains an excellent collection of the actual equipment and apparatus used by him. One room of the house was his laboratory, and in one corner of the room the visitor will see a fume hood. Of course, a servant was required to work the bellows that supplied the draft, but it is interesting to note that safety was a consideration in laboratory work even in the late eighteenth century.

The fact that working with chemicals is an inherently hazardous occupation must surely have been known to the earliest chemists and to the alchemists before them. Such knowledge was probably dearly bought at the expense of illness and occasionally even death.

Until recently, safety was not strongly enforced in academic training of chemists, although it is now widely recognized as a necessary component of academic laboratory work. However, in most colleges and universities it is still not treated in depth, or as a separate subject, but is covered by individual instructors as part of the academic course they are teaching.

The result is that chemists who have recently graduated vary considerably in the extent and depth of their training in laboratory safety. In addition, many laboratories today have numerous nondegreed "technician" workers, many of whom have had only a minimal education in chemistry. These people cannot be expected to recognize the hazards that the materials they work with present, nor to know how to determine these hazards or obtain information about them.

Safety in industrial laboratories may be considered a subtopic under the general heading of industrial safety. And the watershed year for industrial safety in the United States was 1911. Prior to that year, programs dealing with industrial safety were almost nonexistent. A few companies prior to 1911 accepted the moral responsibility to provide safety gear and equipment and to study how manufacturing accidents could be reduced, but not many.

In 1911 the first workman's compensation law was passed. Before that year all states handled industrial injury under the common law employment-at-will concept. When sued for injury or death, employers could almost always escape penalty. In 1911 Wisconsin passed a workman's compensation law that withstood constitutional challenge, and by 1947 all states had some form of workman's compensation.

After 1911 industrial management found itself in the position of having to pay for injuries incurred on the job. It did not require very acute managerial expertise to recognize that it was financially preferable to prevent accidents rather than pay for injuries. This gave rise to the recognition of safety as a legitimate concern of managers and the beginning of an organized safety movement, with books, journals, professional safety officers, and professional safety organizations. In the early years the concentration was on remedying hazardous physical conditions, and statistics show a large drop in serious injuries over the next twenty years.

However, by 1931 the drop in the injury rate had begun to level off and only marginal improvements were being made. That same year saw the publication of H. W. Heinrich's *Industrial Accident Prevention*, a book that was to cause an upheaval in the safety field. The author advanced the revolutionary suggestion that unsafe acts, rather than unsafe conditions, were the cause of a large percentage of industrial accidents. This idea made so much sense that it was accepted almost immediately and has influenced safety thinking ever since. With this philosophy, a whole new concept of accident prevention emerged, and again the injury rate decreased.

For another thirty years steady progress was made in the reduction of industrial injuries, but by 1961 another plateau had been reached. People began to wonder why progress was not continuing. As was common in the sixties, an attempt was made to resolve the situation by passing a law. The Occupational Safety and Health Act, passed in 1970, established the Occupational Safety and Health Administration (OSHA) as a sort of watchdog agency over industrial safety.

OSHA's initial efforts were a return to the apparatus and equipment philosophy that things, not people, cause accidents. However, recent years have seen more emphasis on the establishment of performance standards, which rely on management action. OSHA operates under the close attention and oversight of three major constituencies, which are often in adversarial relationship with each other: Congress, industry, and organized labor. A further obstacle is the technical complexity of many of the OSHA regulations, which are often poorly understood by members of those three groups. It is no wonder that progress in industrial safety often appears to be slow.

To the credit of OSHA administrators, the search for safety standards has continued. One standard was published in 1985, which, together with another that was proposed in 1986, will have a profound effect on the management of general safety in chemical laboratories. In addition, the Hazardous and Solid Wastes Amendments to the Resource Conservation and Recovery Act of 1976 were signed into law in 1984. With this legislation, Congress directed the Environmental Protection Agency (EPA) to bring small-quantity generators of hazardous waste under its system of regulation.

Among the many regulatory actions of OSHA and the EPA, the above are considered to be the most important for chemical laboratory managers. Condensations of the OSHA Hazard Communication Standard (29 CFR 1910.1200), OSHA Occupational Exposure to Toxic Substances in Laboratories (FR/vol. 51, no. 142/July 24, 1986), and Small Quantity Generator Hazardous Waste Rules (RCRA, 40 CFR 261.5) are given below. The Superfund Amendment and Reauthorization Act (SARA) will also impinge on chemical laboratory management in the near future. Title 3 of SARA is considered in chapter 3.

LEGAL ASPECTS OF MANAGING LABORATORY SAFETY

Occupational Safety and Health Administration (OSHA)

While the effectiveness of OSHA in reducing occupational injury may be arguable, there is no doubt that the OSH Act has had a strong influence on the management of safety in the industrial environment, including laboratories:

- OSHA fixed the responsibility for a safe working environment clearly on the employer.
- OSHA provided a mechanism whereby employees could "blow the whistle" on employers who condoned or encouraged unsafe practices, without fear of losing their jobs.

- By means of actual or threatened inspections and fines, OSHA focused the attention of many managers on the safety of the workplaces for which they were responsible.

Although OSHA's original efforts at regulation were concerned mostly with the "nuts and bolts" approach to safety, as mentioned above, in recent years the agency has begun to tackle the human dimensions of the safety problem. One of the first of these efforts was the promulgation of the OSHA Hazard Communication Standard. This standard, adopted in 1983, became effective in part in November 1985, with complete enforcement beginning in May 1986.

The objective of this standard is to make available to employees information on the chemical hazards they may be exposed to in the workplace and to train them in methods of minimizing exposure and injury. The standard initially applies only to employees of manufacturers, distributors, and importers of chemicals, but laboratories are specifically included, although exempt from some of the requirements.

The basic requirements of the standard, as applied to laboratories, are that the employer must post lists of all hazardous chemicals used in the laboratory, maintain a collection of Material Safety Data Sheets on these chemicals, maintain a written hazard communication program, and inform and train employees on the nature of the hazards, means of obtaining information concerning them, and methods of monitoring them and protecting themselves. Although these are the only technical requirements of the standard, the prudent manager will apply all of the standard to his or her laboratory operations. A list of topics covered in the standard is given below. The complete text is given in appendix 1.

OSHA Hazard Communication Standard (29 CFR 1910.1200): Main Topics

- Scope and application
- Hazard characterization

    Flammables, oxidizers, pyrophorics
    Explosives
    Organic peroxides
    Unstable materials
    Water-reactive materials
    Compressed gases

- Health-hazard characterization

- Written hazard communication program
- Labeling
- Material Safety Data Sheets
- Employee information and training

In addition, in 1986 OSHA proposed a standard, Occupational Exposure to Toxic Substances in Laboratories, that would be applicable to all laboratories. It is expected that this standard, or some modified version of it, will be finally promulgated in late 1987 at the earliest. The following is a list of topics covered in this proposed standard. (See appendix 2 for the text of the original 1986 proposal.)

Occupational Exposure to Toxic Substances in Laboratories: Main Topics in Proposed OSHA Rule (FR/vol. 51, no. 142/July 24, 1986/p. 26660)

- Employers to whom standard applies
- Permissible exposure limits
- Requirement for written chemical-hygiene plan
- Topics in written plan: standard operating procedures for working with toxic substances
- Revision and updating of plan
- Requirement for respirators
- Record keeping of employee medical consultations and examinations

Environmental Protection Agency (EPA)

While OSHA has responsibility for regulating workplace safety and health, the EPA is concerned with monitoring and protecting the environment. With the passage of the Resource Conservation and Recovery Act (RCRA) in 1976, the EPA was directed to assume responsibility for the regulation of the generation, transportation, and disposal of hazardous waste.

Laboratories, of course, are generators of hazardous waste (as defined by RCRA). Initially, laboratories were exempt from most of the regulations because they came under the definition of "small generators": organizations producing less than 1,000 kilograms (2,200 pounds) of hazardous waste per month. However, in 1984 Congress directed the EPA to bring the small-quantity hazardous waste generators into the system, and defined small quantities as 100 to 1,000 kilograms per month. These regulations went into effect on September 22, 1986. Many laboratories fall under this small-quantity designation, since only 30 gallons of waste solvent would weigh close to 220 pounds. The following is a summary of the topics covered in this document.

Summary of Small Quantity Hazardous Waste Generator Rules—Resource Conservation and Recovery Act (40 CFR 261.5, 45FR 76623, 46FR 27476)

A hazardous waste is one that, if not disposed of properly, could cause personal injury or damage to land, air, or water. There are two broad categories of hazardous waste: those listed in RCRA regulations and those that are hazardous because of their characteristics. The characteristics that define hazard are ignitability, corrosivity, reactivity, or failure to pass the EPA extractive procedure (EP) test or the proposed Toxicity Characteristic Leaching Procedure.

A small generator disposes of between 100 and 1,000 kilograms of hazardous waste per month, or more than one kilogram of acutely hazardous waste per month.

A small generator must take certain actions:

1. Obtain a USEPA Identification Number
2. If hazardous waste is managed on-site:
   - Comply with storage time, quantity, and handling requirements for containers and tanks
   - Obtain a storage, treatment, or disposal permit
   - Take precautions against accidents and be prepared to handle them properly if they occur
3. If hazardous waste is shipped off-site:
   - Choose a hauler and disposal facility that have USEPA identification numbers
   - Package and label waste for shipping
   - Prepare a hazardous waste manifest

Four important aspects of managing laboratory hazardous waste properly follow:

1. Reduce the amount of hazardous waste
2. Conduct self-inspections
3. Cooperate with local and state authorities
4. Call the state hazardous waste authority with any questions

Judicial Concerns

Mention was made above of the concept of "employment-at-will." This is the legal term for the concept of employment as a contract between employer and employee that can be terminated by either party at will, with no need for explanation. In other words, an employer may fire an employee without being required to justify his action. An employee may also quit without explanation, but it is safe to say that this probably happens much less frequently. This concept dominated labor law until the rise of the labor unions in the last century. Union contracts generally contain specific conditions under which employment may be terminated, as well as procedures for protesting unjust termination.

However, non-unionized employees, for example, most laboratory workers, were still governed by the employment-at-will concept until the early 1980s. With the civil rights movement and the passage of laws forbidding discrimination against persons because of race, sex, religion, or age, the employment-at-will concept has been steadily eroded. In its place there has emerged the idea that a person who has been hired is entitled to that job as long as he or she performs acceptably and the job exists.

The impact of this on safety management in the laboratory and elsewhere has been twofold. There is no doubt that employees have been discharged in the past for being overly concerned about safety by managers who were not sufficiently concerned. This is no longer a viable option to these managers, even if they are able to keep OSHA at bay.

On the other hand, one of the problems of safety management can be recalcitrant employees who object to specific safety procedures. In these cases, the idea of termination can no longer be used effectively to enforce safety regulations without the management safeguards of prior warnings, documentation, and so forth.

Another recent trend in judicial theory that affects safety management is the increasing tendency in the courts to hold managers *personally* responsible for the safety of their subordinates. In a landmark ruling in 1985, two plant managers were convicted of manslaughter for insufficient training and not warning a worker of the dangers of cyanide solutions (Occupational Health and Safety 1984).

SOCIETAL CONCERNS

The past thirty years have seen rising public concern over the hazards that chemicals pose to human health and the environment. Hardly a day goes by without another horror story in the news media: a new carcinogen discovered, a highway or railroad spill of dangerous chemicals, polluted wells, problems in siting hazardous waste landfills, and the like. Love Canal, Times Beach,

Bhopal, Three Mile Island, and Chernobyl conjure up images of a technological society gone awry, threatening the health and existence of its citizens. This has given rise to a widespread popular "chemophobia"—an unreasonable fear of "chemicals."

This fear is most prevalent among those who have little technical knowledge and little appreciation of the actual hazards involved, or of the benefits that chemicals and chemistry offer to the modern world. Nevertheless, the general ferment over chemicals has undoubtedly raised the sensitivity of chemists and laboratory workers to the hazards of their work. This sensitivity in turn affects workers' morale, and the respect they hold for their managers and the organization for which they work. If there is a perception of lack of concern for safety on the part of management, it is safe to say that quality of work and productivity will suffer.

RESPONSIBILITY FOR LABORATORY SAFETY

Safety is obviously a legitimate concern and required function of laboratory management. It may be worthwhile to examine the concept of "responsibility" for safety, and where it lies.

It is a general principle of good management that responsibility accompanies authority, and, indeed, that one without the other can easily lead to disaster. Authority implies the ability or power to command behavior of persons or to allocate resources, and the person with authority must also be responsible for the effective use of personnel and other resources. It should also be remembered that the reverse is true: the person given responsibility must also be given authority or his or her responsibility is compromised.

With this consideration in mind, let us examine the idea of responsibility for safety at three different levels of personnel in the laboratory.

Senior Management

The definition of senior management will vary with the size and nature of the organization of which the laboratory is a part. In general we are talking about the next level of management above the laboratory supervisor who is overseeing the work on a day-to-day basis. Titles such as president, owner, technical vice-president, plant manager, or director of research and development (R&D) describe this level of management.

Senior management's responsibility for laboratory safety includes but is not limited to the following:

- Senior management must have firm commitment to safety and communicate this commitment to all personnel in no uncertain terms, preferably in a written document. Laboratory personnel should be in no doubt that senior management wants a safe working environment, with workers who comply with safety regulations and have a sincere concern for environmental protection.
- Senior management must allocate the personnel and financial resources to provide a safe working environment, safe working practices, and safe disposal of hazardous waste.
- Senior management must delegate responsibility for safety to supervisory management and staff personnel (such as a safety officer) in a clear and unambiguous manner, and hold them accountable for those areas to which their responsibility pertains.
- Senior management holds the ultimate authority for enforcement of safety, that is, the termination or possibility of termination of personnel who willfully violate safety regulations.

Supervisory Personnel

Since senior management cannot effectively monitor the safety of day-to-day laboratory operations, this responsibility is delegated to the laboratory supervisors. Supervisors' responsibilities include but are not limited to the following:

- Assume responsibility for safe working conditions and procedures affecting all personnel reporting to them or working in the areas they supervise.
- Teach employees safety regulations and methods of personal protection.
- Recommend improvements in safety to senior management, either directly or through the safety officer.
- Investigate accidents in their areas of supervision and suggest methods of avoiding such accidents in the future.
- Motivate subordinates to work safely and discipline those who willfully disregard safety.

Individual Personnel

One might expect that the individual laboratory worker would be the one most concerned about safety, since it is his or her life, limbs, or livelihood that are at risk; nevertheless, people are often careless. It is reasonable, however, to expect workers to be responsible for safety, even if it is necessary to make compliance with safety regulations a condition of continued employment. Individual employees should:

- Comply with safety regulations established by laboratory management.
- Bring safety concerns to the attention of their supervisor or safety officer.
- Suggest improvements in operations or equipment for safety purposes.
- Report violations of safety regulations to supervisors. Although this may go against the grain of many people, workers should realize that safety violations endanger not only the violators, but also others in their vicinity, and may destroy the facility if a fire or explosion should occur.

2

Organizing for Safety

Although the ultimate responsibility for safety in an organization must rest with the highest levels of management, it is obvious that top managers cannot spare the time to supervise all operations personally to ensure safety. The responsibility for safety must be delegated to appropriately designated personnel.

The way in which this is done in laboratories will depend to a large extent on the position of the laboratory in the organization, and to a certain degree on the philosophy of management. Laboratories, which are in essence information-generating organizations, serve a variety of functions. The following are some of the more common types of chemical laboratories.

Quality control laboratories are usually part of a manufacturing organization, where they produce information on the quality of raw materials, finished products, and process intermediates.

Technical service laboratories are generally controlled by a marketing operation and work on customer problems, new end-use applications, and competitive products.

Research and development laboratories are usually geographically separate from manufacturing and marketing, and may be under direct management of corporate headquarters.

Commercial laboratories, often called "independent laboratories," are corporate entities performing testing, R&D, and consultation for a wide variety of clients. Essentially, they represent a group of chemists and/or chemical engineers in "private practice."

Government laboratories provide services that are determined by the purpose of the federal, state, or municipal agency that supports them.

Academic laboratories offer the twin functions of teaching and independent R&D.

It is obvious that the chain of command for safety responsibility/authority may be a very tortuous one, depending on the organization that supports the laboratory. In independent commercial laboratories, the safety officer may report directly to top management. In others, such as the quality control laboratory, he or she may report to a plant safety officer, who in turn reports to the plant manager or to a corporate safety department. In academic laboratories, safety may be the responsibility of each instructor, or safety policy may be dictated by a department head or safety officer, or even by a university-wide safety department.

Although the command channel for safety may be devious, it is of prime importance that it be established and that its existence be perceived by all laboratory personnel. Workers must be aware that management is committed to safety and that they are expected to observe safety regulations regardless of their own opinions of the importance or need for such regulations.

Senior management can communicate the commitment to safety in many ways. A direct, formal, documentary commitment in the form of memos, a policy statement, and approval of a safety manual is only the first step. Such documents tend to be ignored unless followed up by positive action.

One of the most important indications to employees of management's commitment to safety or any other policy is the willingness to allocate financial or human resources to the cause. Time spent in safety meetings, in safety committee meetings, and in demonstrations of safety equipment should not be begrudged as long as it is productive in improving safety. The excuse that "We can't afford it" should never be used to avoid purchasing needed safety equipment. If a job or project cannot be done safely, it should not be done.

Informal communication of safety concerns is a valuable management tool. Senior managers visiting laboratories should make a special effort to obey safety regulations themselves (by wearing safety glasses or goggles, for example) and should be sensitive to safety violations. On-the-spot reprimands (in private) for safety violations from a senior manager can be a powerful motivation toward safety consciousness on the part of employees.

Another motivational tool of management is the annual performance review. If employees are rewarded for safety consciousness and contributions to safe operations and penalized for repeated safety violations, management's commitment becomes obvious. Of course, the ultimate penalty for continued, serious violations of safety regulations is termination. Management should be prepared to take this step, if necessary, since the unsafe employee endangers the life, health, and livelihood of his colleagues as well as himself.

THE SAFETY OFFICER

In many of the organizations described above, the responsibility for safety in the laboratory belongs to an individual who is not a member of the laboratory staff. He or she may be the plant safety officer, a safety engineer, a member of the personnel or human resources department, or a representative of some other corporate entity.

This practice, although common, is not recommended. Many of the safety problems encountered in the laboratory environment and in laboratory operations are specific to the laboratory and easily overlooked or misunderstood by persons without training in laboratory work. For example, the major safety concern in most laboratories is the hazard of working with a wide variety of chemicals, which requires some knowledge of chemistry to be understood properly.

It is recommended that, in all chemical laboratories with more than two or three employees, a "safety officer" who is a member of the laboratory staff should be designated. (In this book we shall use the term "safety officer" to refer to this person. Actual titles used may vary considerably depending on corporate custom or nomenclature. Among the titles used are safety director, safety manager, safety coordinator, and safety engineer.)

Position in the Organization

The safety officer in the laboratory monitors operations for safety, advises laboratory supervision on safety matters, and in general serves as a focus for the safety concerns of the laboratory staff. He or she is not responsible for enforcement of safe operating procedures or regulations. This is the responsibility of laboratory supervisors and senior management, since penalty or reward for behavior lies within their responsibilities.

Choosing the Laboratory Safety Officer

The choice of a safety officer will depend to a large degree on the size of the laboratory and the complexity of its operations. In a large laboratory, the position may require a full-time person, or even a staff of two or more. In this case, some consideration should be given to hiring a safety "professional," a person with considerable training in safety and a member of a professional organization with certification.

In many, possibly most, laboratories, the position of safety officer does not require full-time attention, and a working member of the laboratory staff is designated. The choice of safety officer, however, should not be haphazard.

In some laboratories, the position of safety officer is rotated among the technical staff, on an annual or more frequent basis.

The rationale given for this is that it gives safety experience to a variety of people, and that it brings a variety of backgrounds to the job. While this may have had some validity in the past, in the current atmosphere of government regulation and litigation, the inexperience of some of the designated safety officers makes this a dangerous policy.

The following are some considerations that should be kept in mind when choosing a safety officer.

Personal Characteristics

The safety officer should be a mature, practical-minded individual, bringing a sense of responsibility and integrity to the job, since the safety, health, and livelihood of his or her fellow workers may well depend on how well the job is handled.

He or she must have good communication skills, both oral and written. Today's safety officer may also have to deal with compliance with government regulations, as in "right-to-know" laws, or hazardous waste disposal under RCRA. This implies the ability to read and understand these regulations, or to know where to turn for assistance in interpretation.

The safety officer will be responsible for changing the way things are done in the laboratory, and this will bring opposition from people who resist change (which often means everyone). The safety officer must be able to handle such opposition without giving in and without a feeling of personal rejection. In other words, the job is not for the thin-skinned. By the same token, the safety officer must be able to handle opposition diplomatically without alienating the laboratory staff.

Education

Ideally, the laboratory safety officer should be a degreed chemist because of the frequent need for chemical training in evaluating safety problems. It is often difficult, however, to persuade degreed chemists to take on this responsibility. A number of factors may contribute to this situation. The job of safety officer is not perceived as a route to either technical or managerial advancement. Education of degreed chemists in safety is often minimal, and the responsibilities of safety management may be daunting to some.

One way out of this dilemma may be to designate a chemist with other staff responsibilities, such as quality assurance, as safety officer. In any case, the safety officer should have some chemical training or have access to a professional chemist for backup consultation when needed.

Since not all laboratory safety concerns deal with chemicals, the safety officer should also have some knowledge of physics, particularly of the hazards of electrical circuitry. The safety officer will need to be familiar with microbiology if the laboratory engages in this type of work in addition to chemistry.

Training

Since most colleges and universities do not offer courses or degrees in laboratory safety, most safety officer training, unfortunately, is of the on-the-job variety. There are numerous books that deal with the technical side of chemical laboratory safety (see the bibliography at the end of this volume), and the laboratory budget should allow for the purchase of such references.

In addition, the safety officer should be encouraged to attend seminars and short courses that may be accessible to the laboratory. The National Safety Council in Chicago is a good source of information about short courses. The advantages of attending these sessions lie in the up-to-date information imparted and the exposure to outside sources of information.

Membership in local safety organizations should be encouraged. In many communities, municipal or county-wide safety organizations meet on a regular basis to share safety experience and solutions to safety problems. Attending these meetings will benefit the safety officer both technically and psychologically, since he or she will experience solidarity with persons facing the same problems.

Duties and Responsibilities of the Laboratory Safety Officer

- Serves as chairman of the safety committee
- Formulates safety policy for senior management approval
- Writes the safety manual and periodically updates it as new information or new circumstances warrant
- Oversees the maintenance of general safety equipment—fire extinguishers, safety showers, eyewash fountains, and the like
- Educates personnel in safe procedures, safe operations, and use of personal safety equipment
- Supervises and schedules fire drills and emergency and disaster drills
- Supervises the training and operations of the emergency response team

- Investigates accidents, reports them to senior management, and recommends corrective measures
- Conducts safety audits, reports them, and recommends corrective measures
- Monitors and oversees the disposal of hazardous waste
- Supervises compliance with "right-to-know" laws
- Monitors storage, labeling, and use of hazardous chemicals
- Maintains safety-related files, such as accident reports, information on safety equipment, Material Safety Data Sheets, and lists of hazardous chemicals

THE SAFETY COMMITTEE

In some laboratories the safety program is the sole responsibility of the safety officer. In others, there is no safety officer, but a safety committee, made up of technical staff serving for a limited period of time. In still other laboratories there is both a safety officer and a safety committee. The latter is recommended for the following reasons.

A safety committee that operates alone, on a volunteer or appointed basis, lacks the focus and attention on safety concerns that a designated safety officer can bring to the job. Experience shows that such a committee operates inefficiently at best. Even if a chairman is appointed, he or she tends to regard the job as taking time that could be better spent elsewhere. Meetings that are supposed to be held on a regular basis tend to be forgotten or postponed. Corrective action to prevent accidents or injuries is forgotten or not followed up. Safety audits are postponed, and in general the cause of good laboratory safety suffers.

On the other hand, one could argue that, with a good safety officer, the safety committee is somewhat redundant and is not needed. While this may be true, the safety committee does assist the safety officer in doing a better job for a number of reasons.

First, involving working members of the laboratory staff in the safety function serves to raise the safety consciousness of these persons and, if membership on the committee is rotated, eventually a large proportion of the population is made aware of safety. Second, experience shows that members of the safety committee, by virtue of their daily laboratory work, become aware of safety concerns of their fellow workers and can bring these to the attention of the safety officer and the committee.

Finally, the authority of the committee can be used to back up the safety officer in the promulgation of safe operating procedures. Workers are more likely to accept recommendations of a committee of their peers, than those imposed upon them by an individual, especially if the recommendations are unpalatable.

Choosing the Safety Committee

Members of the safety committee should be appointed by the laboratory supervisor(s) with the agreement of the safety officer. Membership on the committee should be representative of the laboratory as a whole and not limited to one class of employee. Nondegreed "technicians" should be represented, as well as degreed chemists. Maintenance and clerical workers should be included, especially as they may have the opportunity to work in the laboratory or handle hazardous chemicals.

The size of the committee will depend on the size of the laboratory. If the laboratory is divided into departments or groups, one or two members from each group can be chosen.

Membership on the committee, with some exceptions, should be for a limited period of time, such as one year or six months. The time period should be long enough to allow everyone to become familiar with the workings of the committee and to contribute to its functions. Membership on the committee should overlap so that experienced members are always in the majority.

The safety officer should chair the committee and be a permanent member. Other persons with some background in safety or health-related fields may be considered for permanent membership. Examples would be a Certified Industrial Hygienist, company nurse, or an employee who is a member of a volunteer fire company.

Safety Committee Meetings

The safety committee should meet regularly, preferably on a monthly basis. More frequent meetings are not generally required, but meeting less frequently may result in postponement of needed safety action. The safety officer should have the authority to call nonscheduled meetings to address problems requiring immediate attention.

Prior to the meeting the safety officer should prepare an agenda of topics he or she will be presenting to the committee for discussion or action. This agenda is distributed to the members before the meeting, along with minutes of the last meeting. This will serve to alert the members of the upcoming meeting, enabling them to adjust work schedules accordingly, and perhaps stimulating them to provide their own topics for discussion.

Committee meetings should seldom require more than an

hour's time. The safety officer, as chairman, should make an effort to keep the meeting on schedule and moving at a moderate pace, without stifling needed discussion. If topics arise that require more information than is available, they should be tabled until the next meeting and the safety officer should attempt to obtain the information in the interim, rather than take time with unproductive speculation.

The meeting should begin with a reading of the last meeting's minutes, followed by corrections if needed. The safety officer can then report any progress or new information on items that were tabled, left open or unresolved. The prepared agenda can then be discussed, followed by a poll of the members regarding any additional items they feel should be brought to the attention of the committee.

Minutes of the meeting should be taken by a designated member of the committee other than the safety officer. They should be typed and distributed to the members as described above, after approval by the safety officer. We recommend that a copy of the minutes be posted on the bulletin boards for the information of the laboratory workers and to make them aware of the topics of concern to the committee. Items in the minutes concerning individuals involved in accidents or injuries should be edited so that individuals are not mentioned by name.

Duties and Responsibilities of the Safety Committee

The safety committee's function in general terms is to give "advice and consent" to the safety officer. By acting as representatives of their fellow workers, committee members can keep the safety officer informed of safety problems that the employees are concerned about. Committee members can also help to disseminate information regarding new rules, regulations, safety equipment, and so on.

Some of the duties and responsibilities of the committee are as follows:

- To serve as monitors of compliance to safe operating procedures in the laboratory
- To make known possible hazards not covered by existing rules and regulations
- To assist the safety officer in promoting safety in the laboratory
- To assist the safety officer in evaluating the effect of proposed regulations on laboratory operations and on the personnel involved

- To assist the safety officer in conducting safety audits in the laboratory
- To assist the safety officer in formulating safety policy and rules and regulations

THE EMERGENCY RESPONSE TEAM

One of the major principles of good management for laboratory safety is expressed in the motto of the Boy Scouts of America: "Be prepared." Although most laboratories spend considerable time and money on means of preventing accidents and injuries, all too often little or nothing is done regarding the problem of handling emergencies after they occur, and they *will* occur in even the best laboratories.

The emergency response team (ERT) is one answer to this situation. The team should be a group of employees with special training in responding to at least the major emergencies that can occur in the laboratory, namely, fire, explosion, accidental injury, sudden illness, and spills of hazardous chemicals.

With the passage of the Superfund Amendment and Reauthorization Act (SARA), the ERT must assume both an internal and external responsibility. In other words, the ERT is not only responsible for cleanup of the workplace, but also for protecting the environment against accidental pollution from spilled chemicals.

SARA establishes a comprehensive *community* right-to-know program. Under this program, state commissions are required to establish emergency planning districts to facilitate the preparation and implementation of emergency plans. Local emergency planning committees in each district must complete an emergency plan, and these plans must be reviewed thereafter on an annual basis. Among the numerous reporting requirements for facilities are the filing of emergency and hazardous chemical inventory forms, and toxic chemical release forms. The text of Title 3, Subpart A, SARA is given in appendix 3.

The team should be under the control of the safety officer or, in his or her absence, the captain of the team. Team members should be carefully chosen. They should be stable individuals, not likely to crack under stress, and potentially long-term employees, since it will be necessary to invest in their training.

The next chapter provides guidelines on preparing for emergencies and discusses the role of the emergency response team.

3

Preparing for Emergencies

An emergency is defined by *Webster's New Collegiate Dictionary* as "an unforeseen combination of circumstances or the resulting state that calls for immediate action." In a laboratory emergency, immediate action is usually needed to prevent or minimize human injury or loss of property. The major objective of every safety program, of course, is to *prevent* emergencies, to establish procedures and have equipment available so that such situations will not occur.

Unfortunately no program seems to be comprehensive enough to guarantee that emergencies will not occur. In the Bhopal, Three Mile Island, and Chernobyl installations, a great deal of attention was paid to preventing emergencies, which nevertheless did occur. It would appear that in any technological system mistakes will happen, and accidents will occur.

The chemical laboratory is an inherently hazardous environment. Workers are surrounded by hazardous materials, and the potential for accident and injury is always present. It is wise to assume then that accidents will happen in the laboratory, and intelligent management requires that steps be taken to prepare an appropriate response beforehand. Strangely enough, this type of preparation is often the most neglected aspect of the laboratory safety program. The remainder of this chapter is concerned with the three major classes of emergencies that are likely to occur in the laboratory, as well as prevention guidelines and responses that will minimize injury and loss of property.

FIRE

Prevention of Fires

Control of Flammable Materials

One of the best means of preventing fires in the laboratory is to establish rules for using flammable materials. This is a management option that is often neglected.

Storage of Flammables

Large quantities of flammable materials, especially liquids, should not be stored in or near the laboratory. Depending on the nature of the work done in the laboratory, a limit should be set on the total amount of flammable liquids permitted at any given time. Laboratory supervisors should be made aware of these limits and take steps to enforce them.

If it is necessary to order large quantities of flammable liquids, these materials should be stored at some distance from the laboratory, preferably in a separate shed or building. Employees can then transport small amounts of these liquids to the laboratory as needed.

Use of Flammables

The use of flammable materials in the laboratory should be carefully monitored by laboratory supervisors. Workers should be encouraged to use flammable liquids in fume hoods as much as possible, and discouraged from working with flammable liquids in open containers on the laboratory bench. Bottles of flammable liquids should be kept tightly stoppered, except when being used.

Control of Sources of Ignition

Smoking should be prohibited in all laboratory areas, as well as other potential fire areas such as the solvent storage shed. Gas burners, such as the formerly ubiquitous Bunsen or Meeker burners, should be limited to those applications where they are absolutely necessary. There are very few applications for these burners that cannot be adequately handled by hotplates or heating mantles. Glassware repair stations should be isolated from other laboratory operations.

Most electrical equipment designed for laboratory use will contain spark-proof electrical contacts and motors. However, many laboratories use equipment such as food processors, electric drills, and domestic-type refrigerators. Designed for nonlaboratory use, these devices frequently are a source of sparks that could cause fires.

Another electrical spark source that is frequently overlooked is electrical tools or welding devices brought into the laboratory by maintenance workers or "outside contractor" personnel. Laboratory supervisors should be aware of this hazard and remove flammable materials before the work is started.

Emergency Response to Fire

The Local Fire Department

A fire in the laboratory is, unfortunately, one of the most common emergencies that can arise, due to the prevalence of flammable materials, especially liquids, and the need to heat them with various potential sources of ignition. Because of the potential for fires, it is prudent for laboratory management to make contact with the local fire department. Fire department personnel, both professional and volunteer, will usually be very cooperative in planning to prevent and respond to fires.

Fire department personnel can be very useful in planning the storage of flammable materials, placing various classes of fire extinguishers, setting up personnel evacuation procedures, and training personnel in emergency procedures. They may also be useful in emergencies other than fires, especially in response to hazardous chemical spills and even personal injuries, since in many communities the fire department also offers ambulance services.

Another source of similar professional assistance may be the laboratory's fire insurance underwriter. Additional information is also available from the National Fire Prevention Association in Boston.

Fire Extinguishers

Fire extinguishers in the laboratory are the first line of defense in containing and controlling fire. Management must give careful consideration to choosing the appropriate type of extinguishers, to placing and maintaining them properly, and to training personnel in the use of extinguishers.

There are four principal kinds of fire extinguishers, classified according to the type of fire they are designed to extinguish.

Type A extinguishers are to be used only on fires of "combustible" materials, such as wood, paper, and similar materials. This type of extinguisher is totally unsuitable for laboratory use, since it is ineffective for highly flammable materials like organic liquids, or on electrical fires. However, they are often used in offices.

Type B extinguishers are designed to be used on organic liquids and other highly flammable materials.

Type C extinguishers are to be used on fires of electrical origin.

Type D extinguishers are designed for use on highly reactive metals or metal compounds.

"Multiple type" fire extinguishers, such as Type ABC, are to be preferred for most laboratory use. Type A should never be used by itself, and Type D is necessary only if the particular compounds for which it is designed are used in the laboratory.

Type ABC extinguishers generally contain "dry powder" as the extinguishing material. While useful for all common laboratory fires, the powder residue left after use can seriously damage delicate instruments or computers. Laboratories containing these kinds of equipment should have "Halon" type ABC extinguishers, which leave no residue.

Extinguishers come in various sizes. The very small sizes are effective only for very small fires, while the largest may be too heavy for persons of small or slight stature to use. Extinguishers of 10 to 15 pounds are most commonly used in laboratories.

Placement of Extinguishers

Every separate room in a laboratory should have at least one fire extinguisher. Large rooms should have more than one, on the general principle that an extinguisher should be readily available to all workers in the laboratory.

Extinguishers should be mounted on a wall near a door or other exit. They should be placed high enough to be easily seen, but not so high as to prevent a short person from easily lifting one down. OSHA regulations require that a sign be mounted above or near the extinguisher, identifying it. Both the sign and the extinguisher should be readily visible from all points in the room. Care should be taken that portable or permanently mounted equipment does not block access to the extinguisher.

Maintenance of Extinguishers

Extinguishers with a permanently mounted pressure gauge are to be preferred. These can be easily checked on a monthly basis to ensure that there has been no loss of pressure. Otherwise, the extinguisher must be weighed regularly to check that it is still operable. Every extinguisher should carry a tag recording the date of the last check and the signature of the person who performed the check.

Extinguishers that have lost pressure must be recharged. Companies that sell and service fire extinguishers will recharge extinguishers at a nominal cost. They will also perform the "hydrostatic pressure" test, which must be done on each extinguisher every five years.

Training of Personnel

Every laboratory worker should receive training in the use of the fire extinguisher. Although lectures, slide presentations, movies, or videotapes provide information regarding fire extinguisher use, actual "hands-on" experience with an extinguisher is invaluable and should be a part of the training. The noise and force of an extinguisher in action can be rather frightening to a person who has never before experienced it.

Many local fire departments are willing to conduct training sessions at low cost for companies within their jurisdiction, including "hands-on" demonstrations on controlled fires outdoors.

In the lecture part of the training session, two important points must be emphasized. These are:

1. Do not stand too close to the fire. When pointed at a burning liquid, the force of the gas from the extinguisher can spread the flames over adjoining walls and ceilings. A distance of fifteen to twenty feet is usually adequate.
2. Once an extinguisher has been used, for no matter how brief a time, it must be recharged. After the seal is broken, the remaining gas in the extinguisher will leak out in a matter of hours. The used extinguisher should be returned to the safety officer for recharging, and the safety officer should have a replacement spare for the laboratory to use.

It is important to emphasize this point. Employees may be tempted to use an extinguisher to put out a small fire and replace the extinguisher on the mount, without bothering to report the incident. When this happens, the lab may go for a month or more with an inoperative extinguisher, even though frequent checks are made.

The training of an employee in the use of fire extinguishers should be documented in his or her personnel file, and should take place within a short time of hiring. Periodic retraining of employees is also desirable.

Announcing the Fire

Every laboratory should have one or more systems to alert all employees in the building to the fact that a fire has started. In large laboratories, or small ones that are part of a large organization, this is often accomplished by a system of fire alarms that is activated by pulling a lever or pushing a button. The rings of the alarm are usually indicative of the location of the fire. There are alarm systems that can simultaneously inform the local fire department.

Some laboratories rely on a public address system, usually accessed through the telephone. A "page" button is held down while an employee announces that a fire has occurred and gives its location. In some telephone systems it is necessary to notify the switchboard operator, who then activates the public address system and makes the announcement. In some laboratories a nonelectrical system such as a hand-operated siren or compressed gas siren is used as a backup for situations where electrical power is lost.

In any case, the first response of an employee to a fire should be to alert his or her fellow employees to the fire and describe its location. The next step should be to notify the fire department.

The latter step is often neglected out of fear of embarrassment at calling out the fire department for a small fire that may well be out by the time the fire truck arrives. However, most fire department personnel say they would rather find the fire already extinguished than be notified only after it has become difficult to control.

Evacuation Procedure

Every laboratory should prepare for the possibility of fire by training all employees in a well-thought-out evacuation procedure. Every exit from the laboratory or from the building in which it is situated is required by OSHA regulation to be clearly designated. It is also prudent to mark doors that lead to a cul-de-sac as "Not an Exit."

Employees should be trained in the "walk, do not run" procedure for evacuation. Running encourages panic and may lead to blockage of exits. Once outside the building every employee should go to an assigned area and report to his or her supervisor. This ensures that the supervisor will be aware of any employee who has not evacuated the building. Supervisors should have a checklist to be certain that no employee is overlooked.

Fire drills, held at least every six months, are recommended to train employees in evacuation and overall response to a fire.

Emergency Response Team

The emergency response team has definite responsibilities with regard to fires. The team, or a subunit, sometimes called the "fire brigade," does not immediately evacuate. Rather, each member of the team picks up the nearest fire extinguisher and carries it to the scene of the fire. Many fires are too large to be put out by a single extinguisher, which is generally good for only twenty to thirty seconds of use. Multiple extinguishers may be able to do the job, especially if wielded by trained personnel.

If they are not able to extinguish the fire, the team members return to their designated workplaces, where they check to be sure that all personnel have evacuated. Special consideration is given to rest rooms and other areas (such as walk-in refrigerators) where persons may not have heard the fire announcement.

INJURY

Prevention

Personal injury is probably the most common emergency encountered in the laboratory. Severity of injuries ranges from the minor cut or burn to life-threatening electrical shocks or ingestion of toxic materials. Prevention of injuries should be of highest priority on the laboratory manager's list of safety concerns. It is accomplished mainly through employee training, the use of personal protective equipment, and the establishment of safe operating procedures for laboratory operations.

Glassware Injuries

Cuts from glassware are the most common injury occurring in chemical laboratories. Glass, because of its chemical inertness and high temperature resistance, is still the material of choice for most laboratory apparatus. Unfortunately, its brittleness and slow stress propagation means that it is easily broken.

Laboratory workers should be instructed never to use chipped, scratched, or broken glassware (with sharp edges). Such glassware should be either discarded or repaired. Often a simple fire polishing is all that is needed.

One of the most common glassware injuries occurs when glass tubing is joined to other materials, such as rubber stoppers or plastic tubing. New employees, especially nondegreed technicians, should be carefully instructed in the proper technique of accomplishing such joining of materials.

Employees should be instructed in the difference between "borosilicate" glass (Pyrex or Kimax brand names, for example) and "soft glass" or "soda-lime" glass. Although all glassware designed for laboratory use is of the borosilicate type, bottles, jars, and other vessels not especially designed for laboratory use are often of the soft glass type. Such glassware cannot be exposed to thermal shocks without cracking or breaking and should not be used in applications involving extremes of heat or cold.

Employees should also be instructed in methods of handling glassware used in high pressure or vacuum applications. Under such conditions glass is especially vulnerable to mechanical shock, which can lead to explosion or implosion.

Employees should be encouraged to substitute plasticware for glassware wherever appropriate.

Toxic Chemical Injuries

Prevention of these injuries is also best accomplished through personnel training and strictly enforced safety regulations that should include the following:

- Chemicals that give rise to toxic fumes or vapors should be handled in a fume hood whenever possible. If it is necessary to handle such materials outside of a hood, adequate spot ventilation or appropriate respiratory protection should be used.
- Employees, especially nondegreed technicians, should be carefully instructed in handling materials that can react to give off dangerous fumes or gases, such as sulfides and cyanides.
- Appropriate gloves and other clothing should be used when handling chemicals that are toxic by skin absorption.
- Rules for handling samples and other materials of unknown composition should be established and enforced.
- Rules for disposal of toxic chemicals should be established and enforced.
- Food and beverages should never be consumed in the laboratory, nor stored in the laboratory. Laboratory refrigerators should never be used to store food or beverages intended for consumption.
- Employees should be encouraged to wash their hands before eating, drinking, or smoking and before leaving the laboratory at the end of the day.
- Pipetting chemicals by mouth should *never* be permitted, even if the materials are known to be nontoxic.

Treatment of employees exposed to toxic chemicals often requires specific knowledge of proper first-aid procedures. One of the best sources for this are Material Safety Data Sheets (MSDSs), which the OSHA Hazard Communication Standard now requires chemical manufacturers and importers to supply. Laboratories in the industrial sector must maintain MSDSs on file, but all laboratories should do the same. The MSDS contains information on first-aid procedures specific to the particular chemical it covers.

A source of similar information is the various computer data bases for hazardous substances currently being offered by chemical supply houses and other companies. One problem with such data bases is that it is often difficult to get "on-line" with the host computer during normal business hours.

Burns

Burns, both chemical and thermal, are another common injury in the chemical laboratory. Prevention depends on employee training and provision of proper equipment for operations and for treatment of burns.

Eyewash fountains and safety showers should be provided at accessible locations throughout the laboratory. This equipment should be checked monthly to be sure it is working and the check documented on tags attached to the equipment.

Chemical burns are usually caused by contact of corrosive acids or alkalis with the skin, particularly strong acids. Employees should be instructed in the proper methods of transporting, handling, and transferring such materials.

Employees should also be instructed in methods of handling hot objects, using tongs, thermally resistant gloves, and so on. Employees should also be instructed in procedures for using heating devices such as hotplates, muffle furnaces, and heating mantles.

One of the most dangerous accidents that can occur is an employee's clothing catching fire. Workers should be instructed to drop to the floor and roll to smother the flames, or to flood with water under a safety shower, if one is nearby.

Electrical Shock

Laboratories use a wide variety of electrical equipment. A few simple protective measures can go a long way in preventing this type of injury.

All electrical equipment should use three-prong plugs, where the third prong is the "ground" plug. Although all equipment manufactured for laboratory use currently follows the three-prong standard, older equipment and devices such as electric mixers, food processors, and other home appliances often do not. Such equipment should be modified to provide for grounding.

Although most building codes require three-hole (grounded) sockets, older buildings may not have them. Such buildings should be rewired and three-hole sockets provided. The use of three-hole to two-hole "adapters" is not recommended because they often do not give a good ground. Sockets in new buildings should be checked to be sure that they are, in fact, grounded.

Some laboratories have both 220 volt and 110 volt sockets. The 220 volt line sockets should be so designed that a 110 volt device cannot be accidentally inserted. Once again, this is standard practice in new laboratories, but may not be the case in older buildings.

Power cords on equipment should be frequently inspected for damage. Damaged cords should be replaced immediately.

Employees should be instructed not to attempt repairs on electrical equipment. These should only be done by a qualified technician or electrician.

Emergency Response

Medical Assistance

Every laboratory should have a source of medical assistance in case of employee injury. In laboratories that are part of a large organization, this is usually provided by the presence of a "dispensary" with a part-time or full-time nurse or doctor. However, accidents cannot be guaranteed to occur only when medical staff is at hand. It is well to prepare for all emergencies by having a working understanding with a local hospital, clinic, or medical center that can offer twenty-four-hour service. This is especially important for the laboratory that does not have "in-house" medical resources.

First Aid

The immediate response to an injury in the laboratory should be "first aid." For minor cuts, burns, or bruises, this may be self-applied from a first-aid kit. Such kits, which are available from safety equipment supply houses, contain bandages, surgical gauze, burn salve, pain-killers such as aspirin or acetominophen, and disinfectants.

First-aid kits should be available to all laboratory personnel. It is important that kits be inspected periodically and their contents replenished as needed. Documentation of inspection should be provided by an attached tag or sticker.

More serious accidents, such as those involving arterial bleeding, unconsciousness, or loss of breathing, require more knowledgeable first aid. Every laboratory should have one or more persons on staff who have had formal first-aid training or, as it is often called nowadays, EMT (Emergency Medical Training).

Such training involves learning methods to stop bleeding, artificial resuscitation, cardiopulmonary resuscitation (CPR), and prevention of shock. Persons who have received such training should be members of the emergency response team.

Transportation to Medical Assistance

After first aid, the next step in emergency response should be to get medical assistance for the victim. Often this requires transportation to the hospital, clinic, or medical center.

The cardinal rule here is that the victim should never be allowed to proceed unaccompanied. Victims have bled to death on the way to the hospital. Delayed shock syndrome can occur, causing a drastic drop in blood pressure and blood volume, with severe consequences if not treated.

CHEMICAL SPILLS

Prevention

The best preventive measure for chemical spills is to allow only small quantities of chemicals and solvents to be kept in the laboratory. Large drums or multiple bottles of such chemicals should be stored away from the laboratory in a separate building or shed. Small quantities can then be brought into the laboratory as needed. Since spills seldom involve more than one container, spills in the laboratory will consist of about 2½ liters of chemical at the most.

Employees should be instructed in the proper technique of transporting hazardous chemicals from storage to the laboratory or from one laboratory to another. Special padded or rubber bottle carriers should be provided to prevent breakage by accidental striking against walls or floor, and to contain the material if breakage does occur.

Employees should be instructed never to store hazardous chemicals on the floor. They should be kept on low shelves or in cabinets. The shelves should have a "lip" on the forward edge to prevent bottles from slipping off when accidentally struck or moved.

In laboratory operations large bottles of hazardous materials should never be picked up by the lid or by the molded glass ring sometimes provided. Two hands should always be used, with one under the bottle and the other around the neck.

The laboratory should provide special "spill pillows," available from most safety supply houses. These are porous, inert plastic cushions filled with fumed silica, which is highly absorbent. Spill pillows are useful for picking up all chemicals except hydrofluoric acid.

Emergency Response

Small chemical spills generally do not constitute a laboratory emergency. They are easily mopped up, absorbed by paper or cloth, and disposed of. However, spills of chemicals of more than one liter volume may be considered an emergency, especially if the chemical is hazardous. The emergency response team should be given special training in handling chemical spills.

Volatile, flammable organic solvents present the hazards of fire or explosion. All sources of heat or ignition should be immediately extinguished. Maximum ventilation should be ensured by opening windows and outside doors if possible. Fume hoods should be turned on, if not already operating.

Personnel involved in cleanup should wear respiratory protection if the material is sufficiently volatile to give rise to dangerous concentrations. The MSDS should be consulted for relevant information. Plastic, slip-on covers for shoes and plastic gloves should also be used.

Spilled material should be absorbed by spill pillows, rags, paper, or other absorbent material. The absorbed material should be placed in waste drums or other containers for disposal by accepted, legal methods. The residue may be mopped up with soap or detergent and water and the waste water collected and disposed of in the sewer if local sewer authorities permit.

Consulting the MSDS file for particular chemicals will assist the team in cleaning up spills, since this type of information is provided for each hazardous substance.

Spills of corrosive liquids such as acids and alkalis require somewhat different cleanup techniques. Respiratory protection is usually not needed, but disposable shoe covering and plastic gloves will probably be useful. Spill pillows are useful for absorbing these materials (except for hydrofluoric acid), but organic materials such as rags or paper should be avoided. Absorbed materials should be collected in waste containers and disposed of. The residue, in the case of acid, should be treated with sodium bicarbonate until release of carbon dioxide ceases, after which it may be flushed down the drain. Similarly, weak acids, such as acetic or dilute hydrochloric, may be used to neutralize alkalis, although care must be taken not to over-treat.

Mercury presents a special hazard when spilled in the laboratory. Because of its high density and surface tension and low viscosity, spilled mercury tends to form very small droplets that bounce and roll into small crevices, under benches and chair legs, and so forth. Some of these droplets are small enough to be difficult to see. Special spill kits for mercury are available. These consist of a vacuum device for picking up large droplets, sponges for small droplets, and a powder that amalgamates whatever mercury is left.

REPORTING AND DOCUMENTING ACCIDENTS AND INJURIES

All accidents, whether they result in injury or not, should be reported. Employees should report them to their supervisor, who should report them to the safety officer. It is important to emphasize to employees that even minor accidents must be reported. There is a natural tendency to resist reporting minor cuts and bruises out of fear or perhaps embarrassment, or because a person doesn't want to be considered a weakling.

However, there are two good reasons for insisting that minor accidents be reported. The first is that minor injuries can sometimes lead to more serious complications, such as an infection or symptoms that only appear at a later time. In this case it may be important, for insurance purposes, to be able to establish that the injury was work-related. The other reason is that minor accidents may be symptomatic of a situation that could result in a major accident, if not corrected by management action.

Safety professionals argue that an accident almost always indicates a breakdown or oversight in the management system. The accident investigation process must therefore determine not only the immediate cause of the accident, but also those managerial factors that may have contributed to it.

To be effective, the investigation process must be perceived as having a strong commitment from top management. This can be accomplished by high management visibility during the investigation, emphasis on uncovering the causal factors in the accident rather than assigning blame, and decisive action to correct deficiencies, including those traceable to management. In other words, management must support the investigation and act on the results.

It is important that employees understand that the purpose of reporting and documenting accidents is not to affix blame or responsibility for an unfortunate occurrence, but rather to uncover the cause of the accident so that future accidents may be prevented. In discussing the accident, managers and safety officers should be careful to project a sympathetic attitude and avoid expressions of blame or accusation.

The type of accident report that is used will depend on whether or not the laboratory is exempt from the OSHA recordkeeping requirements. This will depend on the Standard Industrial Classification (SIC) number of the organization of which the laboratory is a part. The local OSHA office may be consulted to determine the SIC code for your laboratory and its exempt or nonexempt status. Those laboratories that are not exempt must report injuries on OSHA form 200. Injuries reported must be from a work accident or an exposure in the workplace environment resulting in death, illness, or an injury involving medical treatment, loss of consciousness, restriction of work or motion, or transfer to another job. A detailed report of the injuries must be made on form 101 or an equivalent form containing the same information. Some equivalent forms are: Workman's Compensation Injury Report forms from the state labor department, insurance claim forms, American National Standards Institute (ANSI) form ANSI 216.2 (1968), or the form suggested by the National Safety Council (see National Safety Council 1983).

If the laboratory is classified as exempt according to the SIC code, a program for documentation of accidents and injuries must nevertheless be developed. When an accident is reported, the safety officer should initiate documentation of the incident. The investigation report should include:

- Company name and specific address of the location where the accident occurred.
- Time and date of the accident.
- Person(s) involved, including Social Security numbers and length of employment.
- Names of eyewitnesses, if any, and Social Security numbers. If the accident occurred off-premises, addresses should also be obtained.
- A complete and accurate description of the accident, including usual occupation of the injured employee, occupation at the time of the accident, description of the accident, and supervision, if any, at the time of the accident.
- Whether medical assistance was necessary, what assistance was given, and by whom administered (name and address).

- A statement by the supervisor of the cause(s) of the accident, and a description of the sequence of events leading to it.
- A statement by the supervisor concerning steps taken to prevent similar accidents from occurring.
- Recommendation(s) by the safety officer for preventive measures.

A typical accident report form is shown on pages 35–36. Part I of the form should be filled out by the safety officer, after personal discussion with the people involved in the accident. It is very important that the description of the accident be clear and complete. Rather than vague statements such as "employee was hurt on the thumb," the description should be as specific as possible: "Employee suffered a cut on the fleshy part of the right thumb." Sometimes it requires a considerable amount of digging on the part of the safety officer to determine exactly what occurred and why.

Part II of the form gives the employee a chance to review the safety officer's statement and make any corrections he or she thinks are necessary. The signature on this part can be important from a legal standpoint if later litigation occurs.

Part III is completed by the supervisor, who is asked to state the cause of the accident, the reason the cause existed, and measures taken to prevent similar accidents in the future.

Part IV is completed by the safety officer, after consultation with the safety committee and the supervisor. Every effort should be made to identify causes of accidents and to make recommendations for prevention of future occurrences. It is often very easy to find that the accident "just happened" and that nothing can be done to prevent future accidents. This is not an acceptable response for either Part III or Part IV of the report. Recommendations may cover new equipment, new procedures, or better training of employees.

Completed accident reports should be circulated to senior managers, who are asked to initial them as proof of receipt. Copies should be placed in the employee's personnel file and in the safety officer's file.

Accident Report Form
XYZ Laboratory

Part I (to be filled out by Safety Officer)

Date of accident:
Time of accident: AM PM

Person(s) involved:
Supervisor:
Eyewitnesses (if any):

Description of accident:

Treatment or medical assistance received:

Signed (Safety Officer): Date:

Part II (to be filled out by person(s) involved in the accident)

Is the above information correct? Yes No

If not please indicate corrections:

Signed: Date:

continued

Part III (to be filled out by Supervisor)

Describe the cause of the accident (see Part V):

What is the reason this cause existed?

Signed (Supervisor): Date:

Part IV (to be filled out by Safety Officer)

Recommendations for preventing future accidents of this type:

Signed (Safety Officer): Date:

Part V (for supervisors): Although we want you to describe the cause of the accident in your own words, the following list may assist you in defining the cause.

Unsafe Conditions: Unsafe procedures, poor illumination, poor ventilation, improper shielding, defective equipment, poor design of equipment, poor housekeeping

Poor Supervision: Lack of training, safe operating procedures not enforced, protective equipment not provided, proper or safe tools not provided

Unsafe Acts: Operating without authority, haste, shortcuts, protective equipment not used, horseplay, unsafe position or posture, bypassing safety devices, using unsafe or improper equipment, taking undue risk, poor technique, failure to secure

Unsafe Personal Factors: Lack of knowledge, inattention, improper attitude, poor physical condition, illness

4

The Safety Manual

A good safety program involves establishing performance standards for workers' behavior in the workplace, assigning responsibility for their enforcement, and communicating these standards and assignments to the employees. While oral communication of the safety standards may be important in some cases, it does not take the place of a written safety manual, which has several advantages.

The manual will ensure uniformity in compliance to the safety requirements and, if clearly written, will leave little room for misinterpretation. It can be used to explain the reasons for the various safe operating procedures, thus reinforcing workers' desire to comply. This is particularly important in the case of nondegreed technicians, who may not understand the technical rationale for many of the rules. A written manual can also serve as documentation that employees have been instructed in the details of the safety program in case of litigation or government investigation.

Since the laboratory safety officer is the person responsible for establishing most of the rules and for monitoring compliance, he or she should have responsibility for writing the manual and for updating it as necessary. (The actual writing may be done by someone else under the supervision of the safety officer.)

Every laboratory has unique operations involving safety precautions, and it is therefore not possible to find a "fill-in-the-blanks" type of manual that will be suitable for any given laboratory. As an example, the evacuation scheme to be followed in case of fire must correspond to the structure of the building the laboratory is in. This means that the manual must be tailored to fit the individual laboratory.

WRITING THE MANUAL

The safety manual, like any document, should be written with a specific target audience or reader in mind. In this case the reader should be assumed to be a person with minimum technical education and experience who would be hired for laboratory work. In many laboratories this will be a nondegreed technician, who may be assumed to have had a high school course in chemistry.

The writer should not assume any technical competence greater than a high school chemistry student would have. Jargon should be avoided, unless it is first defined. "PCBs," for example, while understood by most chemists to stand for "polychlorinated biphenyl compounds," is not necessarily so understood by the general populace. A college student of one of the authors misunderstood "ppm" to mean "pounds per minute," instead of "parts per million."

An effort should be made to explain the reasoning behind all safety regulations, and not to simply present them as a set of prescribed "rules." This will help to motivate employees to obey the requirements. The writing should, above all, be clear and unambiguous.

Before beginning the writing of the manual, the safety officer should make a list of all safety practices currently being observed in the laboratory, based on discussions with laboratory supervisors and workers. In addition to this list, the safety officer should add those practices that he or she feels ought to be implemented. Agreement with these proposed practices may be sought at this time, or after the draft version of the manual is circulated.

The physical production of the manual should reflect the fact that frequent revision or additions may be necessary, and that the quantity needed will be small. The use of a computer with word processing and disk storage is recommended to make revision easy. Means of reproduction may be multilith (if the computer's printer can handle it) or photocopying. The manual should be written in sections, with pages numbered by section and page within the section (that is, page II-5 would be the fifth page in section II). This enables easy expansion of a given section without having to renumber the whole book. A loose-leaf or similar binder should be used so that pages may easily be added or replaced when revisions are made.

When the manual is completed, a copy should be given to each employee, and to each new employee on the first day of employment or before. To ensure that the manual is read, his or her immediate supervisor should get together with the employee to discuss the major points. The last page may be a form requiring a signature, to verify that the employee has read the manual and agrees to abide by the safe operating procedures as a condition of continued employment. The signed form should be returned to the supervisor and kept in the employee's personnel file. This strategy may not have value as a legal tool in cases of suit by an employee because of injury, but it does serve to enforce the concept that obedience to safety regulations is not a matter of personal choice.

SUGGESTED FORMAT AND TOPICS

Although it is necessary for each laboratory to produce its own safety manual, for reasons given above, we include here some suggestions for a format, and a list of topics that should be considered for inclusion.

Title Page

This page should include, in addition to the title, the date of issuance, authorship, and the name of the laboratory. It may also include space for authorization by one or more members of senior management.

Table of Contents

A table of contents is absolutely necessary if the manual is to be used as a ready reference by the workers in the laboratory. An index is a useful feature, but requires considerable effort to prepare.

Part I: General Considerations

The following topics may be included in this part.

Objectives. A statement should be made of the corporate philosophy toward safety and the objectives of the safety program. This should incorporate a commitment to maintaining a safe environment for workers and visitors; providing the necessary facilities, manpower, and equipment for safety; and protecting the environment from hazardous chemicals and waste.

Responsibility for safety. This section should delineate the responsibilities of various levels of supervision for achieving the objectives of the program.

Senior or "top" management oversees the program and is responsible for allocating the necessary human and financial resources. This management level also has ultimate responsibility for enforcement of the safety regulations, including discharge of persons who willfully disregard them.

Supervisory managers have responsibility for ensuring the safety of personnel operating in areas under their control, and the safety of persons reporting to them when operating in other areas. Supervisory managers also are responsible for training personnel in the safety aspects of laboratory operations in areas under their control.

Nonsupervisory personnel are responsible for learning the

safety regulations, abiding by them, and reporting any unsafe acts or conditions to supervisors.

The responsibilities of the safety officer and safety committee are described in chapter 2.

Part II: Specific Safety Rules and Regulations

The following topics may be included in this section.

Emergency procedures. This should describe what to do in the case of fire, accidents, chemical spills, and other emergencies.

Safe working environment. Regulations on housekeeping, food and drink consumption, smoking, odors, and other environmental considerations may be covered.

Personal safety equipment. Eye protection, clothing, gloves, hard hats, rubber and plastic aprons, and respiratory protection are suitable topics.

General safety equipment. Fire extinguishers, safety showers, eyewash fountains, and first-aid kits may be covered in this section.

Ventilation. Hoods and local ventilation ducts should be discussed.

Disposal of waste chemicals. General requirements of the Resource Conservation and Recovery Act, Subtitle C, must be followed, in particular Parts 261 through 265.

Transportation and transferring chemicals. Proper methods for carrying bottles, for pipetting (never by mouth), and for mixing chemicals are suitable topics.

Compressed gas cylinders. Safety rules for storage, transportation, and use must be given.

Storage of chemicals. The need for labels showing date received and date opened should be discussed.

Glassware safety. Fire polishing, prohibition of use of cracked or chipped glassware, the proper method of connecting rubber and glass tubing, and disposal of broken glassware may be covered in this section.

Heating and cooling devices. Safety considerations in the use of gas burners, hot plates, heating mantles, drying ovens, muffle furnaces, refrigerators, and cooling mixtures using dry ice should be covered.

Electrical equipment. Avoiding electrical shock, proper grounding of equipment, and use of three-hole sockets and three-pronged plugs are types of subjects that should be covered.

Miscellaneous. Safety in field operations, sample handling, automotive safety, and other matters not previously covered may be included here.

This list of topics is not intended to be comprehensive or applicable to every situation, but to serve as a guide for a manual for most laboratories. Many laboratories have operations, such as the use of radioactive materials, lasers, or X-rays, that will require special consideration. Microbiology laboratories, which are often part of the same organization as chemical laboratories, require additional safe operating procedures.

OSHA CHEMICAL HYGIENE STANDARD

The Occupational Safety and Health Administration has proposed a new "chemical hygiene" standard for laboratories under Occupational Exposure to Toxic Substances in Laboratories. The proposed standard is not in final form, since comments have been requested and public hearings must be held before promulgation. One of the major requirements of the standard is that management must have a written "Chemical Hygiene Plan" (see appendix 2 for the text of the proposed standard).

It should be pointed out that the OSHA standard covers safety only in the sense of controlling exposure to hazardous chemicals, and not physical hazards like fires or electrical hazards. However, it would seem prudent to cover all safety regulations in one document. Appendix A of the OSHA standard suggests *Prudent Practices for Handling Hazardous Chemicals in Laboratories* by the National Research Council, published in 1981, available from the National Academy Press, 2101 Constitution Avenue, NW, Washington, D.C. 20410, as a comprehensive guide.

5

Personal Protection

The use of personal protective clothing and equipment is of major importance in minimizing accidents and injuries in the laboratory. (Nonpersonal items such as fire extinguishers and safety showers, which are available for everyone to use in emergency situations, are discussed in chapter 3.) Virtually all laboratories have some form of personal protective devices available, but what is often not recognized is the need for management leadership in the proper use of such equipment. Employees must not only be trained in its proper use, but monitored to see that the training is effective.

Senior or top management's responsibility in this area is to allocate the necessary funds to purchase equipment, and to show a commitment to its use. Laboratory supervisors must determine the need for such equipment, instruct their subordinates, and evaluate the effectiveness of both the equipment and the training. The safety officer should provide independent monitoring of equipment and use, assist or conduct training sessions, and serve as a resource for employees in obtaining information on equipment and how to use it.

The proper use of protective clothing and equipment involves technical questions touching on the sciences of chemistry, physics, and biology. The field is changing rapidly as a result of new regulatory requirements, additional research findings, and the development of new materials. The prudent manager therefore will seek expert assistance in deciding questions of equipment and its use. Professionals in the field of safety and industrial hygiene are available, either in-house or as independent consultants. Suppliers of safety equipment and clothing will also be found to be an excellent resource for information in the field. Meetings of the American Society for Testing and Materials (ASTM) Committee F-23 on Protective Clothing are excellent forums for obtaining information on proper clothing.

EYE AND FACE PROTECTION

The eye is the most valued sense organ, and also probably the most vulnerable because of its fragility. Protection of the eyes is therefore probably the most important facet of personal protection. The need for safety glasses in laboratory operations is universally recognized. However, in many laboratories, even though safety regulations require personnel to wear safety glasses, enforcement of the regulation is lax. Management should realize that simply having a safety regulation requiring the use of safety glasses will not remove management's responsibility if an eye injury occurs.

The major problem in enforcing the safety glasses requirement usually occurs with persons who do not normally wear glasses. They feel that the glasses are uncomfortable and a nuisance and will remove them at the earliest opportunity. Such persons should be encouraged to wear them at all times, even in lunchrooms and the like, during the workday, until they become used to them. Of course, complaints about comfort should be referred to a professional optician for correct fitting.

Occasionally, persons whose eyesight may be on the verge of requiring correction may complain that they cannot see well with the plano (uncorrected) glasses supplied, especially when doing close work. This is because the glasses do slightly decrease the amount of light reaching the eye. Such persons should be referred to an optician to determine if a slight correction in the glasses might be beneficial.

Employees should be encouraged to obtain a new prescription before obtaining safety glasses. Unfortunately, the term "safety glasses" may be confusing. In 1972 the Food and Drug Administration (FDA) ruled that all prescription eyeglass and sunglass lenses must be impact resistant. However, such lenses are not the equivalent of industrial safety glasses and should not be used in a laboratory. Industrial safety glasses must pass a more stringent impact test, and frames as well as lenses are impact resistant. Most industrial safety glasses have the designation Z 87 stamped on the frame and have a manufacturer's logo on the lenses. The Z 87 indicates that the glasses conform to the American National Standards Institute (ANSI) standard for safety glasses.

Another management problem in establishing an eye protection program is the prevalence in today's population of contact lenses. Most safety professionals agree that contact lenses should not be worn in a laboratory. One reason is that in the case of a splash the eye's natural reflex is to clamp shut. This may make it very difficult to remove the lens to irrigate the eye, and extensive damage may be done. Another reason is that the plastic lenses are permeable to many of the gases and vapors present in the laboratory, and irritation of the eye results.

Some employees who wear contact lenses claim that they cannot switch to glasses during the workday, because their eyes would not then remain accustomed to the contacts. Others may have vision problems that are better corrected with contacts than with glasses. One solution to this problem is to permit the use of contacts, but only if covered with safety goggles of a type designed to be used with contacts. These are goggles with no openings that permit entry of gas or vapor. Unfortunately they tend to be uncomfortable, especially in a humid atmosphere, and few employees voluntarily accept this solution.

Job applicants should be questioned regarding their use of contact lenses and informed of the laboratory's policy before a job offer is made and accepted.

An effective eye protection program is a necessity for every laboratory. The following guidelines will be found helpful in establishing such a program.

- The safety officer should designate those operations in which eye protection is required. Careful consideration should be given to the designation of areas. For example, it may be decided to exclude all offices from the need for eye protection. If so, be sure that there are no offices that contain small laboratories or are close enough to a laboratory to be subject to hazard.
- One hundred percent participation should be required. This should include not only laboratory workers, but all visitors and temporary workers (outside contractors or maintenance personnel, for example).
- Signs such as the following should be posted at the entrances to all designated areas: "All Personnel and Visitors Must Wear Eye Protection."
- In many laboratories safety glasses are provided at no cost to employees. The glasses should be properly fitted by a professional. Optical companies will perform this service and instruct employees in proper maintenance. Every effort should be made to ensure that new employees are provided with proper eye protection as soon as possible, and temporary protection (goggles or plastic glasses) provided in the interim.
- Eyewash fountains should be readily available to all personnel in the laboratory.
- Wherever possible, eye hazards should be controlled at the source. For example, safety guards against splashing or flying objects, or enclosures to confine dusts, mists, or vapors can be installed.

- Contact lenses without appropriate eyewear should not be worn in the laboratory.
- Supervisors and other personnel should be aware that safety glasses are not the ultimate in eye protection; these individuals should enforce the use of face shields and safety shields in areas or operations where appropriate. For example, in glassware cleaning operations where large quantities of acid dichromate are used, face shields should be mandatory.
- Safety glasses, face shields, and other eye protection devices should be periodically inspected for scratches, fogging, or crazing. If the devices are so damaged that they reduce vision, they should be replaced immediately.

EAR PROTECTION

Protection of hearing is not a concern in many laboratories. Personnel will often accept the annoyance of noise from instruments, pumps, grinders, mechanical sieves, and high speed devices as part of the operation since there is no hearing problem apparent at the time. Hearing loss can occur after long-term exposure, however. It is incumbent on managers to determine the amount of noise to which personnel are exposed on a long-term basis.

Noise levels should be measured by a certified industrial hygienist, who will use either a sound level meter or a dosimeter. A sound level meter is a hand-held device that will read noise level in dBAs. A dosimeter is worn by an employee during his or her workday to give a continuous measure of noise exposure. Readings from the instruments can be compared with standards on a time-weighted average basis to indicate potential for hearing damage. Hearing protection should be mandatory for an exposure exceeding 85 decibels on an eight-hour time-weighted basis.

Several courses of action are available for reducing or avoiding noise. Noise can be reduced by enclosures, acoustical barriers, or other engineering controls. Administrative action can help reduce noise exposure. Signs should be posted in areas where hearing protection is required. Jobs can also be rotated among personnel or exposure time can be shortened.

Personal protective equipment use is a major and effective means of reducing noise exposure. For the laboratory, the use of earplugs or earmuffs is recommended.

Earplugs may be custom molded or purchased as a one-size-fits-all type. The latter generally are disposable. Earplugs have the advantage of not restricting head clearance. They also can be used without disturbing hairstyle or eyeglasses, and they do not interfere with other personal protective equipment. Some disadvantages of earplugs are that they can work loose because of jaw action, they have a short service life, and they can be easily lost because of their small size.

Earmuffs are generally of rigid plastic and contain an acoustical barrier material. A cushion cup and a soft inner lining make them comfortable to wear. The cup should feel snug and surround the entire ear. Earmuffs are awkward to use and their use in the laboratory should be carefully considered. If earmuffs are used and become damaged or deformed in any way they must be replaced immediately.

NOSE AND MOUTH PROTECTION

Respirators are not generally used on a routine basis in laboratories, but special circumstances, such as cleaning up after a spill or entering a contaminated space, may require their use. Fieldworkers in industrial hygiene or hazardous waste laboratories may require respirators. The laboratory emergency response team should be trained in respirator use and should have various types of respirators available.

OSHA has established that respiratory protection is needed in four types of hazardous atmospheres.

1. Atmospheres containing gaseous contaminants, including gases and vapors of volatile liquids
2. Atmospheres containing particulate contaminants, including dusts, mists, and fumes—that is, particles or droplets small enough to remain suspended in air over a sustained period and be inhaled
3. Atmospheres containing combined contaminants, comprised of both particulate and gaseous contaminants
4. Atmospheres that are oxygen-deficient, which occur when the oxygen in the air is consumed by chemical reaction such as fire or when oxygen is displaced by another gas

The existence of a hazardous atmosphere and the magnitude of the hazard are most accurately determined by means of air sampling and analysis. If this cannot be done in-house, an outside company offering such services may be used (for example, one that offers industrial hygiene services). In emergency situations such as a chemical spill or rescue operation, where time is important, reliance must be placed on whatever information is available and a "worst case" assumption made.

Choosing the proper respirator is a decision best made by a person with some expertise. If the safety officer lacks the necessary knowledge, he or she may consult an industrial hygienist. The local fire department may also be a good source of information. Material Safety Data Sheets, which are required to supply information on spill cleanup procedures, should be consulted.

The law requires that in nonemergency situations, feasible engineering controls and/or administrative controls must be instituted to reduce employee exposure to acceptable levels. If these controls are inadequate or not feasible, respiratory protection must be provided.

There are four basic types of respirators:

1. Air-purifying devices that filter out particulate contaminants
2. Air-purifiers that contain cartridges or cannisters for the removal of organic vapors and gases
3. Atmosphere-supplying respirators that exclude workplace air and supply respirable air through a hose from a stationary source located outside of the workplace
4. Self-contained breathing apparatus (SCBA) with an air supply that may last from three minutes to four hours, depending on the device

Respirators must always be used in the context of a respiratory protection program. Such a program should include:

- Written operating procedures governing respirator selection and use.
- Air sampling and evaluation of contaminants.
- Medical examinations to determine the fitness of employees to wear respirators. Some persons, because of respiratory problems that may not affect normal operations, should not be asked to wear a respirator.
- Training and fitting for all employees who may be required to use respirators. Male employees with thick mustaches and/or beards are often difficult or impossible to fit properly with respirators.
- Information for employees on how to select proper respirators for the job at hand.
- A program for storage, cleaning, inspecting, and repackag-

ing respirators. Respirators should be inspected before and after each use, sanitized after each use, and packed and stored so as to maintain proper condition.
- An auditing program to be conducted at appropriate intervals.

PROTECTIVE CLOTHING

While the need for protective clothing in the laboratory is usually confined to gloves, aprons, and shoes, situations may arise where more extensive protection is desirable. Examples of such situations are chemical spill cleanup operations, the handling of carcinogens or other high chronic toxicity materials, frequent sampling in a manufacturing operation, or the handling of hazardous waste. Manufacturing operations where personnel are exposed to hazardous material in large quantities and for longer periods of time have much more stringent requirements for protective clothing. This has given rise to a great deal of research on "barrier properties" of materials used for protective clothing and a great diversity of protective apparel. Information pertinent to almost any problem in this area is available from safety professionals, industrial hygienists, and companies selling safety equipment and clothing.

The Barrier Concept

A barrier is any material or combination of materials used to isolate parts of the human body from a particular hazard or challenge chemical. The rapid growth of the chemical industry in this century soon gave rise to concern over exposure of personnel to hazardous chemicals in large quantities. Many of these chemicals posed special problems due to chronic toxicity; that is, toxic effects that arise from long-term exposure at relatively low levels. The original barrier fabrics were simply oilskins or fabrics covered with natural rubber (rainwear). While providing more protection against splashes than cotton work clothing, the old barrier fabrics were generally ineffective.

Improved protection has resulted from the development of synthetic fibers such as nylon and polyester for stronger substrates, and synthetic elastomers like butyl and neoprene rubbers for markedly improved barrier properties. Lamination processes enabled the use of more than one elastomer in the same fabric, and recent developments using nonwoven fabrics have lowered costs to the point where disposable garments are often practical. As an example, one new composite fabric combines the chemical inertness of Teflon with the good strength and flammability properties of Nomex.

Some pertinent chemical properties of barriers are:

- Resistance to degradation, that is, any change in physical properties resulting from contact with a challenge chemical
- Resistance to penetration, or the flow of a challenge chemical through closures, pinholes, seams, or other imperfections in a barrier fabric
- Resistance to permeation, or penetration of a barrier fabric by a challenge chemical on a molecular scale

A permeation test measures two parameters: breakthrough time and permeation rate. The most commonly used parameter is the breakthrough time, or the time between initial contact of a chemical with the outside of a barrier fabric, and the appearance of the chemical on the inside of the fabric. Breakthrough times generally range from fractions of a second to days, although tests are not usually run for more than eight hours.

Early tests for barrier fabrics were often based on degradation of the fabric after a timed immersion in the challenge chemical. These tests, although still used by some vendors, can be misleading, because some fabrics that are not degraded are actually permeated rather quickly. Conversely, other barrier/challenge combinations may show rapid and extensive degradation and still remain impermeable over long periods of time.

Most protective clothing manufacturers will provide customers with tables of permeation data using their barrier fabrics versus a variety of common chemicals. Computer data bases of barrier properties are also appearing on the market (for example, *CPCbase—The Chemical Protective Clothing Data Base* 1985).

Barrier chemical properties will vary widely depending on the nature of both the barrier material and the challenge chemical. In addition, barrier properties can vary with processing conditions used in the manufacture of the barrier material. Permeation properties, for example, will differ, due to manufacturing technique, from one manufacturer to another and from lot to lot for a given manufacturer. Therefore, when specifying protective clothing, the manufacturer's name and catalog number should be included along with the type of barrier fabric.

Barrier Physical Properties

Some of the important physical properties of barrier materials are flexibility, abrasion resistance, tear resistance, wet or dry grip (of gloves), puncture resistance, antistatic properties, aging properties, flammability, and thermal resistance. Test methods that are available for each of these properties may be used to compare one barrier versus another.

An evaluation of the integrity of an entire glove or suit is usually a better criterion of quality than tests on individual components. It is important to evaluate sample clothing on the specific job for which it was designed, since performance will also depend on such physical components as seams, closures, and face shields in total encapsulating suits. Like the weak link in a chain, poor seams or closures and needle holes can easily damage the overall effectiveness of protective garments.

Decontamination of Protective Clothing

Protective clothing used in the laboratory will become contaminated. Decontamination between wearings is therefore necessary unless the clothing is disposable. Recent investigations (Garland 1986) have indicated a number of pitfalls that should be borne in mind:

- Surface cleaning does not remove a contaminant held within the matrix of the barrier fabric.
- Significant amounts of a contaminant can enter a barrier matrix and represent a potential hazard.
- Chemicals that enter a barrier matrix, whether as a challenge chemical or a cleaning solvent, have an adverse effect on permeation properties.

Selecting Protective Clothing

The selection process is best handled by a qualified industrial hygienist, who will assess the risks of exposure, define the chemical resistance and physical properties required, and match them to available clothing. Some factors that should be considered are:

- Quantities and concentrations of chemicals used on the job.
- Toxicity, corrosivity, and allowable exposure limits for each chemical.
- Physical state (solid, liquid, gas, or particulate) of the challenge chemical and most probable route of body entry (inhalation, skin absorption, or ingestion).
- The likelihood of chemical contact on the job—is contact probable or unavoidable?

Chemicals for which protection will be required should be

listed, together with the expected clothing wear time for each task. Permeation data for available barriers and/or clothing should be obtained and garments selected that have breakthrough times greater than the expected wear times. Attention should also be paid to the required physical properties and to matching them with the required barrier chemical properties. Material Saftey Data Sheets may give information on protective clothing, but it is often too general to serve as a criterion for selection. Such recommendations as "use impervious rubber or plastic materials" are not very helpful.

It is obvious that selection of whole-body protective clothing requires a great deal of effort. A minimum inventory of protective clothing should be maintained, but it should be sufficient to do the job safely without harmful exposure. Where barrier properties permit, use one type of clothing for as many different jobs as possible.

Documentation of the initial selection process and subsequent revisions is important. For each item of protective clothing there should be a rationale behind the selection, design specifications from the manufacturer, and performance specifications from on-the-job experience.

Laboratory Clothing

While the foregoing considerations apply to those special circumstances where thorough protection from chemical contamination is required, in most laboratory operations only minimal body protection is needed. In most laboratories the basic requirement is for a lab coat, supplemented as necessary by additional clothing.

Lab Coats

In most laboratories, employees are required to wear lab coats. By itself, however, this garment is not adequate protection against chemical contamination and especially not against chronic exposure hazards. However, the lab coat will give some protection in case of splashes, at least to the extent of minimizing the amount of chemical contacting the skin.

Three lab coats should be provided per employee, to allow for laundering and accidental contamination. Laundering should be handled by a firm that specializes in decontaminating chemically soiled garments and can handle the wastes generated, if these services are not available in-house.

Laboratories that do not provide lab coats or require employees to provide their own should carefully rethink this decision.

Not only do lab coats provide some protection, but they promote an atmosphere of professionalism and enhance employee morale (in spite of protestations to the contrary). Employees should be cautioned against wearing shorts or other clothing that exposes large areas of skin, since this may compound hazard exposure from spills or splashes.

Aprons

Plastic or rubber aprons are often provided for short-term protection when, for example, employees handle large quantities of corrosive or toxic chemicals. If this is a common or recurring operation, the selection process described above should be followed.

Gloves

Gloves are probably the most common form of protective clothing used in the laboratory. Fortunately, gloves that offer protection from virtually all types of exposure hazards are available. The selection process described above should be applied to choosing gloves, especially in those cases where long-term or recurrent exposure is expected, or where especially toxic materials are to be handled. Laboratory safety books list barrier materials for gloves and various challenge chemicals (see, for example, National Research Council 1981).

Physical properties such as dexterity and grip (wet and dry) are important in glove selection. Chemical properties should be selected based on permeation data. When available in required barrier materials with sufficiently long breakthrough times, disposable gloves should always be used. Nitrile-dipped cotton gloves should be considered in lieu of leather gloves for handling glassware and general laboratory equipment. The nitrile-dipped cotton gloves are reported to have excellent cut and abrasion resistance and cost less than leather.

Thermally resistant gloves are necessary for handling hot objects and exceptionally cold material. Asbestos gloves are no longer available or considered a viable option because of the carcinogenic hazard. Fortunately, acceptable substitutes such as fiberglass and Zetek have been found.

Shoes

In normal laboratory work special shoes are not necessary. However, workers should be cautioned against wearing sandals, open-toed shoes, or cloth shoes. Hard leather shoes offer at least minimum protection in case of spills or the dropping of heavy objects. Leather shoes readily absorb organic liquids, however, and should be discarded if they become contaminated. Employees handling heavy objects should be provided with steel-toed safety shoes.

For spill cleanup and other special situations, boots made from a variety of chemical barriers can be used. Some of the more common barrier materials are neoprene, butyl rubber, natural rubber, and polyvinyl chloride (PVC). Inexpensive, disposable booties are also available.

Special Applications

It is impossible to anticipate the need for protective clothing or devices for all possible types of laboratory work. Work with radioactive materials, X rays, or lasers may require special devices and protective clothing. A large variety of products have been developed for this purpose. If you cannot find what you need, contact suppliers and manufacturers, who may be able to custom design something to meet your needs.

6

Safety in Handling Hazardous Chemicals

Without doubt the major cause of injury in the chemical laboratory arises from the necessity to work with hazardous "chemicals." It should be borne in mind that the word "chemical" used as a noun is almost impossible to define (at least as it is used in the contemporary world). The chemist will be quick to point out that everything in the universe is composed of "chemicals" in the sense in which it is used in scientific discourse. On the other hand, to the layman, "chemical" is used to denote any solid, liquid, or gas that is man-made. Many natural materials are among the deadliest of compounds, but are not perceived as being in the same class as "chemicals."

In the laboratory we tend to restrict the word to reagents, solvents, reactants, and products used or produced in the laboratory. However, from a safety standpoint, we should recognize that there are other chemical hazards, for example, from samples received in the analytical laboratory or maintenance materials such as cleaners, gasoline, welding rods, and soldering fluxes.

The term "hazardous" is similarly vague. There are, of course, official definitions of the term promulgated by the EPA in connection with RCRA, and by OSHA in connection with workplace industrial hygiene. However, it is not difficult to select chemicals that are easily recognizable as hazardous, but that do not fall within those official definitions. The degree of hazard posed by a given chemical can also vary considerably depending on factors such as concentration, physical state, temperature, as well as user characteristics such as sex, age, or lifestyle.

Therefore, most experts in the field agree that the prudent course is to treat all chemicals as being potentially hazardous, rather than to attempt detailed hazard assessments and create handling methods for each chemical substance. Some general rules to use when handling hazardous chemicals are presented below.

STORAGE OF CHEMICALS

Chemicals are stored in the laboratory in storerooms or "stockrooms," and in the laboratory itself. The following are some suggestions for safety in chemical storerooms.

The storeroom location should be carefully selected. Its atmosphere should not be too cold, hot, moist, or dry. Chemicals should never be exposed to direct sunlight. These requirements rule out, for example, hot, dry attics and cold, damp basements.

The storeroom should not be used for laboratory operations of any kind. It should be the responsibility of one person, the only individual to have access to the chemicals on a regular basis. Other employees may be given access at times when the storeroom clerk is not available. Under no circumstances should an "open" storeroom with access by all personnel be allowed.

Chemicals should be arranged on the storeroom shelves in accordance with the chemicals' reactive properties. Acids should be separated from alkalis, oxidants from reductants, and so on. Explosive chemicals and flammable solvents should be stored separately, preferably in an area blocked off from the remainder of the chemicals. The common practice of storing chemicals alphabetically should not be used. Storage spaces on the shelves should be indexed and a file system provided correlating each chemical with the space where it is stored.

Stored chemicals should be provided with a label indicating the date received and the date the container was first opened. This information should not be scratched on an existing label, but recorded on an added label. An inventory system—preferably computerized—should be implemented. The computer can keep accurate inventory and be programmed to flag chemicals that have been in storage longer than a predetermined time, or that are due to be reordered.

Large quantities of flammable or corrosive chemicals should not be kept in the storeroom if it is located in the laboratory building or any high-occupancy area. They should be stored in an outside shed, and minimal amounts brought into the storeroom as needed.

Storeroom shelves should have an outside lip to prevent containers from accidentally slipping off. Flammable solvents and corrosive liquids should not be stored on high shelves. Glass containers should not be allowed to contact each other. A protective "boot" may be used on alternate glass containers. Another safeguard is to use a catch-pan or protective container for corrosive or flammable liquids.

Adequate, positive ventilation (pressurized above atmospheric pressure) should be provided for the storeroom at all times. Air conditioning and warm air should be provided during the appropriate seasons.

All chemicals in the storeroom should be inspected at least semiannually. It is important to dispose of chemicals that show signs of deterioration or that exceed a predetermined shelf life, as well as containers that no longer have their labels.

Virtually all laboratories store chemicals that are frequently used in the laboratory itself. Many of the above principles also apply in this case. As has been stated previously, however, the total amount of chemicals stored in a laboratory should be kept to a minimum.

The laboratory storage area should also be chosen with some care. The shelves for laboratory storage should be placed away from sources of heat such as heating pipes, radiators, laboratory ovens, and furnaces, and out of direct sunlight.

Laboratory storage shelves should also be provided with a lip to prevent slippage. Corrosive liquids should be placed on low shelves, but never on the floor, except in a closed cabinet. Flammable, volatile liquids should be stored in separate cabinets provided with localized exhaust fans.

Large quantities of corrosive or flammable liquids should never be kept in the laboratory. The quantity allowed in the laboratory should be commensurate with the daily usage.

All chemicals should be labeled with the date received, the date first opened, and the expiration date, if applicable. These dates should not be scratched on the manufacturer's label, as is common practice, since this may obscure important information and damage the label. Printed labels are commercially available that have space allotted for the date received, the date opened, and the expiration date.

In general, chemicals should never be stored in a fume hood, since this may interfere with the proper operation of the hood and may increase the danger or damage in case of fire or explosion.

Laboratory supervisors should inspect all laboratory-stored chemicals at least once each year and see to the disposal of those that are unlabeled, deteriorated, or past their expiration date. Chemicals no longer frequently used should be returned to the storeroom.

Chemicals requiring refrigerator storage should be in containers with screw-cap lids that have Teflon or polyethylene liners. Cork, rubber, or unsealed glass stoppers can leak fumes or allow spillage if the container is toppled.

Highly toxic, carcinogenic, or explosive material should never be stored in a refrigerator used for other chemicals. If necessary, a refrigerator can be designated for one use and no other. In such cases a sign on the door should identify, for example, a "toxics only" refrigerator.

All refrigerators used for chemical storage should be explosion-proof.

WORKING WITH HAZARDOUS CHEMICALS

To list all the safety considerations of working with hazardous materials would be beyond the scope of this book. Many guidelines are substance-specific, and there are thousands of hazardous substances. Other guidelines cover the mechanics of using protective devices and equipment, a topic that concerns safety management primarily as an aspect of employee training. It is prudent to treat all chemicals as if they were hazardous, in keeping with the old precept that it is better to be safe than sorry. The following are general guidelines that reflect this philosophy and address specific safety matters that are often neglected.

Large containers (two to three liters) of chemicals being transported over long distances should always be encased in buckets or special "acid bottle carriers" to avoid accidental breakage. Plastic buckets should generally be used, although metal pails are acceptable for organic solvents. The bucket should be large enough to contain the contents of the bottle.

Within the laboratory, bottles may be carried by hand, but should always be carried by using *both* hands, one underneath the bottle and the other around its neck. Under no circumstances should employees be permitted to carry such bottles by the cap, or by the molded glass handle provided by many vendors.

The general principle of never allowing pipetting by mouth is now widely recognized, but is still violated in some laboratories, especially those with older employees. This rule should be rigorously enforced by managers and supervisors. Under no circumstance should the mouth be used for suction.

As a general rule organic liquids should always be used in a hood. These materials are ordinarily flammable or toxic or both. On the other hand, the hood should never serve as a means of disposing of such liquids by evaporation of large quantities. If such evaporation is necessary, a means of collecting the vapors, such as distillation or scrubbing, should be provided.

All chemical containers should be properly labeled, and employees should be instructed never to use the contents of an unlabeled container. This applies not only to materials in the vendor's container, but also to laboratory-prepared reagents and other solutions or mixtures. The label on such solutions or mixtures should show the identity and concentrations of components, the date of preparation, the expiration date if applicable, and the name of the preparer.

Laboratory clothing that has become contaminated should be immediately removed and laundered before being worn again. Grossly contaminated clothing should be disposed of as hazardous waste.

Waste containers in a laboratory should be taken to a central waste storage area and emptied at least once each week. The waste storage area should be located away from the laboratory, preferably in a separate building or shed.

Work with highly toxic chemicals should be carefully supervised. Any procedure that may generate vapors, dusts, or aerosols of highly toxic compounds should be carried out under a hood. Gloves and other protective clothing should always be used. Hands and arms should be washed immediately upon completion of the work. Containers of highly toxic materials should be kept in trays, and trays should be placed under apparatus used in the work, to contain the material in case of spillage.

Work with carcinogens, mutagens, teratogens, or any other high chronic toxicity material should be supervised with particular care. Some carcinogens have special OSHA rules for handling and record keeping. One problem here is determining which compounds are carcinogenic. The local OSHA office may be contacted for lists of known or suspected carcinogens. The regional EPA office may also be contacted for this information. In addition, chemicals such as polycyclic aromatic hydrocarbons, alkylating agents, chlorinated hydrocarbons, or aromatic amines should be treated as possible carcinogens.

All work with these materials should be carried out in an isolated area with access restricted to those persons aware of the hazards and of the compounds being used. The isolated area should be decontaminated before normal work is resumed. A plan should be prepared for use of the chemicals, decontamination, and waste disposal before work is begun. For example, protective clothing should be removed on leaving the isolated area, conspicuous signs should be placed outside the area, and gloves should always be used in working with high chronic toxicity materials. Special care should be taken to isolate and decontaminate or dispose of waste.

Because of the long latency period between exposure and development of symptoms (sometimes as long as twenty to thirty years), special attention should be given to record keeping. Records should be kept of the amounts and identities of the materials used, dates of use, and the names and Social Security numbers of exposed workers.

Work with explosive materials should be carefully supervised. Although commercial explosives, such as nitroglycerine and TNT, are not common in laboratories, there are relatively common chemicals that are explosive in certain circumstances or that can deteriorate to yield explosive mixtures.

Compressed gases. Compressed gas cylinders present several hazards if not properly handled. They should never be allowed to stand without support, such as chaining to a bench clamp. They should be transported only by hand truck, clamped to it firmly, and capped.

Pressure regulators specified for the gas type should always be used. When the pressure falls to twenty-five pounds square inch gauge, the cylinder should be marked "empty" and deposited in the proper area for return to the manufacturer.

Cylinders containing flammable gases should never be stored or used in the vicinity of open flames or other ignition sources or in the vicinity of cylinders containing compressed oxygen or air.

Most gas cylinders are designed to be used at temperatures up to 50°C (122°F). However, as a general principle, they should not be placed in warm or hot areas such as near furnaces, radiators, and so on.

Perchloric acid. Although the dangers of working with this material are well known, employees, especially technicians, should be carefully instructed in its use. It should be used *only* in a hood designed for its use, constructed of noncombustible material, and provided with means for water washdown. Contact with organic materials should be avoided. Spills, even of diluted acid, should be promptly cleaned up, since dilute acid can evaporate to become concentrated. Perchlorate esters and salts of transition metals are also capable of exploding.

Picric acid. This compound is a high explosive when dry ($<$ 10 percent water). It is normally shipped with 10 to 15 percent water, in which form it is relatively stable (although toxic). The problem is to keep the water content above 10 percent, since below this level picric acid is shock, spark, heat, and often friction sensitive. One insidious problem is that small crystals may become lodged in the threads of the cap and explode when the cap is removed.

One suggested method of handling picric acid is to weigh the container before and after each use. Any loss in weight between uses may be assumed to be water, and the material should be discarded without opening (American Chemical Society 1985b, 44–45).

Picrate salts are very shock sensitive, wet or dry. Therefore, contact with metals, ammonia, or ammonia-containing compounds that can form picrate salts should be avoided.

Peroxide formers. Several common laboratory chemicals present a special hazard because they will form peroxides on exposure to air. The peroxides are shock sensitive and may explode. The most common laboratory solvent of this type is ethyl ether. Containers of ether should be labeled with date received and date opened, and either used up or disposed within one month of opening. Any container of ethyl ether that has stood six months after opening should be treated with extreme care. Although there are various tests that can be performed to detect peroxides, they require manipulation of the material, which may be hazardous.

Other relatively common peroxide formers are: p-dioxane, tetrahydrofuran, isopropyl ether, tetralin, and decalin. Classes of compounds that may form peroxides are: aldehydes, cyclic ketones, ethers, allylic compounds ($CH_2CH{=}CH_{2}-$), and compounds with benzylic hydrogen atoms (American Chemical Society 1985b, 39–41).

One source of hazardous materials that is often overlooked is material of unknown composition received in the laboratory. Such material is commonly received as samples by laboratories engaged in environmental analyses, especially independent, commercial laboratories.

A special isolated reception area should be set aside for these materials. The area should be well ventilated, with at least one large hood in good operating condition, and should contain space for storage of samples prior to analysis. Workers in the area should be provided with protective clothing and gloves and should wear them at all times.

An effort should be made to obtain as much information as possible from the submitter regarding the actual or possible components of the samples. Care should be taken in evaluating this information, however, since those submitting the sample often have insufficient technical knowledge. To minimize disposal problems, agreement should be made before accepting samples that unused portions will be returned.

Another source of hazardous chemicals often not recognized is material used in maintenance of the physical plant, such as cleaning compounds, detergents, disinfectants, gasoline, paint, and motor oil. Although commonly used in nonlaboratory circumstances, these materials can be as hazardous as many laboratory chemicals, and should be handled accordingly. Maintenance personnel and clerical workers, as well as laboratory workers, should receive information and training in handling hazardous materials.

7

Managing Personnel for Safety

If this were a perfect world, there would be little need to persuade laboratory workers to take safety precautions or to use safety equipment, since simple self-preservation would be a sufficient motive. However, experience shows that human beings are seldom rational in their approach to safety. One need only mention the case of car seat belts: some 50 percent of American riders refuse to wear seat belts in spite of their proven safety record and the fact that improved design has minimized the discomfort of using them. State laws have had to be passed requiring seat-belt use.

The truth of the matter seems to be that most human beings have an innate belief in their own immortality and freedom from harm. Although logic dictates that precautions be taken, they are often neglected for convenience or speed of working.

Until fairly recent times the attitude of management was that workers were to be educated in safe practices but following them was the workers' choice. This is no longer the case. Recent court decisions have made clear the principle that management's duties extend to enforcement of safety regulations. In other words, it is not sufficient for a manager simply to inform his employees of safe practices and provide the necessary equipment; he must also plan for safety and motivate his people to work safely. Not only have the courts embraced this philosophy, but many labor unions have also taken this attitude. It therefore behooves the laboratory manager to accept the management of safety as part of his essential function in the organization.

PRINCIPLES OF PERSONNEL MANAGEMENT

The general principles of personnel management can be summed up in three words: communication, consistency, and control. The need for communication is obvious for we cannot expect people to perform acceptably if we do not communicate our notions of acceptable performance. Consistency means fairness and equal application of rules of expected behavior. Control, in this context, means setting up a system or systems of monitoring performance to determine whether performance is acceptable. Let us examine the application of these principles to safety management.

Communication

Borrowing a basic principle from journalism, we aim to communicate the who, what, when, where, and why of safety.

Who? Obviously, all employees who regularly work in the laboratory are expected to observe necessary safety precautions. "Transients," such as visitors, office workers on an errand or passing through, maintenance workers, or outside contractors' employees, must also be considered.

What? What exactly are the safety precautions employees are expected to observe? A safety manual (see chapter 4) will codify safe operating procedures for most cases. However, special training and education in safety are also included under this heading.

When? Obviously, the safety program applies to time spent in the laboratory. A lost-time injury, however, affects laboratory operations whether it occurs on or off the job. Many organizations therefore emphasize safety in the home and automobile as well as at the job.

Where? Again, safe operating procedures apply in the workplace, but many laboratories send employees out in the field or to customers' plants. Safe operating procedures for these occasions should also be communicated.

Why? It is important that employees know the underlying reason for a given safety rule. Some of these are obvious, but many require a technical explanation. Unless employees know why they are being asked to comply with a safety rule, they are likely to disregard it.

A safety manual will go a long way towards communicating these aspects of a laboratory's safety program. However, no manual can incorporate all of the information that employees need in order to work safely. The following are additional means of communicating safety information.

Training Sessions

Oral communication of the most general and important safe operating procedures should be part of every employee's initial orientation session. This should include such topics as the need for eye protection, smoking policy, eating or drinking in the laboratory, and what to do in case of fire. This is a good time to present the laboratory's safety program and to give the new employee a copy of the safety manual. This orientation should occur as soon as possible after an employee is hired, and the fact that orientation was given should be documented and the document placed in the employee's personnel file.

In addition to the orientation session, many organizations hold safety meetings on a regular basis. These are usually held on the department level, with the laboratory supervisor or safety officer presiding. The meetings usually focus on one aspect of safety, either a procedure or a problem, that is of immediate interest to the members of the department. Safety meetings are held at least monthly and sometimes as often as weekly or biweekly.

Written Memos

Safety information may be disseminated by written memo to laboratory employees. A memo may contain information that the safety officer wishes to communicate immediately. If important enough, the memo may have a space for employees to sign, indicating that they have read the memo. Other memos may be posted on the bulletin board.

Safety Committee Minutes

As described previously, minutes of safety committee meetings should be publicly posted to keep employees informed of the topics concerning the committee.

Visual Aids

There are numerous videotapes, slide presentations, and movie films available on many aspects of safety. These aids may be obtained from companies manufacturing safety equipment, laboratory supply houses, safety organizations, or local fire departments. In some cases a small fee is charged, but other presentations are available at no cost. They usually cover a single area of safety, such as fire extinguishers or glassware safety, and can be used to emphasize that particular topic. All such visual aids should be previewed by the safety officer and the safety committee before being presented, to be sure that they do not contain material that is counter to the laboratory's safety program.

Material Safety Data Sheets (MSDS)

All manufacturers, importers, and suppliers of chemicals are now required by law to supply an MSDS with the first shipment of a chemical or whenever the MSDS is updated. The MSDSs contain a wealth of information on a given chemical, including the hazards of overexposure, first-aid procedures, protective equipment, spill containment, and cleanup measures. MSDSs should be kept on file by the safety officer, and employees must also have access to them and know where they are kept. Employees should be instructed on how to read and interpret an MSDS, since much of the information is given in highly technical jargon. The person who does not know what a flash point is or how it is measured cannot easily interpret this number. Instruction concerning MSDSs can be part of the training session mentioned above.

Other Information Aids

There are many other sources of safety information that can be used. Magazine articles on safety can be circulated. Books on safety abound. A separate section of the laboratory's library can be set aside for safety books, or certain books such as *Prudent Practices for Handling Hazardous Chemicals in Laboratories* (National Research Council 1981) may be kept in the laboratory for ready reference. Cartoons dealing with safety may be posted on bulletin boards, and posters dealing with various aspects of safety are available from various sources.

Consistency

Consistency in personnel management means treating employees fairly or, in other words, treating all persons equally. Safety regulations, to be effective, must be perceived by employees to apply to all workers.

The manager who refuses to wear safety glasses or goggles when visiting the laboratory work areas or who neglects to extinguish a cigaret or cigar can effectively sabotage a good safety program. The message he sends to the workers is that he pays only lip service to safety principles.

All visitors should be provided with safety glasses or other acceptable eye protection before being allowed into the laboratory, and smoking in these areas should be prohibited for all persons. Employees of outside contractors such as builders, maintenance workers, and instrument repair technicians should be expected to follow the same safe operating procedures as laboratory employees, especially since some of these people may be spending extended periods of time in the laboratory.

Above all, consistent violation of safety regulations cannot be tolerated, no matter how valuable or essential the employee in question. Preferential treatment will only encourage others to ignore safety and thus degrade the entire safety program.

Control

"Control" as used in management theory relates to a system or systems by which management receives feedback on the success of a program. In the case of safety, the two most effective means of assessing and maintaining control are through accident/injury reports and safety audits.

Accident/Injury Reports

As described in chapter 3, reports of incidents are prepared by the safety officer with the cooperation of the employees involved and their supervisors. These reports, which include suggested means of preventing similar accidents, should be circulated to laboratory management before being filed.

The safety officer may use these reports as the basis of an annual evaluation of the effectiveness of the safety program, in comparison with previous years, and as a way of identifying areas that need additional attention. For example, if an excessive number of glassware injuries has occurred, it may mean that additional education in the use of glassware is indicated. In most laboratories, accidents and injuries are infrequent enough to make statistical analysis meaningless, but common-sense analysis of this type of data can be a valuable aid in improving the safety program.

Safety Audits

A formal audit of safety in the laboratory should be made periodically, preferably on a quarterly basis, and at least annually. The audit should be conducted by the safety officer, preferably accompanied by one or more members of the safety committee. The members of the safety committee will help give credence to the audit team's findings and may also become aware of needed safety improvements that have escaped the safety officer's notice. The audit should be conducted on an "unannounced" basis to ensure that no special precautions are taken to prepare for it.

Various techniques can be used for conducting audits. The audit team's function is to observe the laboratory environment and personnel for conformance to the laboratory's safety program and principles, and to identify any other potentially unsafe conditions or practices. One method is to assign different responsibilities to various members of the team. For example, one member could inspect for electrical safety, another for handling of hazardous chemicals, and a third for use of protective equipment.

Another technique that is often used is the checklist. While it is a quick way to conduct an audit, this method has its disadvantages. It is impossible to cover every item of concern on a checklist. Too often it is assumed that if the laboratory gets good marks, everything is satisfactory, but the audit team may well have overlooked a blatant safety violation because it is not on the list. On the other hand, checklists are useful as memory aids for ensuring that important items are not overlooked or forgotten.

After the audit, the safety officer should prepare a report detailing the findings and recommendations for improvements. The report should be sent to senior management and laboratory supervisors and a copy filed. Within thirty days—or sooner, depending on the urgency—a follow-up inspection should be made by the safety officer to determine the status of any recommended changes. Problems with instituting these changes should be documented and reported to management.

In addition to these formal safety audits, the safety officer may elect to conduct informal audits on his or her own, as time permits or when specific problems may arise. Informal audits usually focus on one specific aspect of safety, such as the status of fire extinguishers, first-aid kits, or safety showers. These audits should also be reported to management, even if no problems are found, to document that the audit was undertaken, when, and by whom.

Job Safety Audits

Another useful technique is the job safety audit (JSA), which is used for jobs that are repetitive and done frequently, or those that may be especially hazardous. A good example would be routine analytical work performed on a large number of samples.

A JSA is a written document that reviews all steps in a procedure, evaluates the potential for hazard or damage in each step, and recommends safe job procedures. It is also an invaluable training aid for instructing personnel in laboratory procedures.

A sample JSA form is shown on page 68. The form should be filled in by the supervisor or safety officer while actually watching the performance of the job. Some points to be followed in developing and using JSAs are:

- Start with the jobs that have the greatest potential for hazard and damage to property.

Job Safety Audit

Job: Date:
Operator: Supervisor:
Department: Laboratory:
Reviewed by: Approved by:
Required/recommended personal protection:

Sequence of Job Steps	Potential Accidents, Hazards, or Damage	Recommended Safe Job Procedure

- Keep JSAs up to date. Review when any changes in material or equipment take place.
- Make sure that all steps in the job are included.
- If an accident occurs, review the JSA to see if any changes should be made.

Supervisors should be required to review all JSAs with employees on a regular basis. A record of the contact should be kept. It can serve as a tool in assessing safety performance of supervisors, as well as a record of employee training.

MOTIVATING SUPERVISORY PERSONNEL

The mechanics of a good safety program, which include purchasing safety equipment, appointing a safety officer, and writing a safety manual, are relatively easy to establish. However, an *effective* safety program involves imbuing all laboratory personnel with a sense of the need for safety in all laboratory operations. We need to teach personnel to "think safety." This means developing in employees a positive attitude that causes them not to simply *react* to accidents and injuries, but to *act* to prevent them.

The key to developing this type of attitude lies with the first-line supervisors in the laboratory. These are the people who are responsible for the work performed in the laboratory on a day-to-day basis and who have overall responsibility for the welfare of their subordinates. Supervisors should:

- Teach safe procedures to subordinates
- Always be alert to unsafe practices, procedures, and situations in the areas for which the supervisors are responsible
- Ensure that "transients" in the laboratory obey safe operating procedures
- Ensure that subordinates who are sent to other areas obey relevant safety procedures
- Evaluate all new operations, including analytical methods or research projects, for safety before they are permitted to be implemented in the laboratory
- Be aware of the hazardous waste being generated, ensure that it is properly disposed of, and take steps to reduce it
- Take innovative measures to improve the safety of personnel for whom the supervisors are responsible

In most cases, laboratory supervisors will not automatically have this type of attitude toward safety when they are first appointed. This aspect of their job is viewed as secondary to the technical and other supervisory duties. It is vital that senior management develop and nurture this kind of safety consciousness in supervisory personnel.

Supervisors must be held responsible for safety problems in their areas. If an employee is injured, or a fire or other accident occurs, the supervisor, after thorough investigation, must be held responsible for corrective action to prevent such injuries or accidents in the future. Note that the supervisor is not necessarily responsible for the injury or accident since he or she may not have been aware that the hazard existed. However, the supervisor may, and probably should, be held accountable for any future injuries or accidents of the same type.

Supervisors who condone or ignore flagrant violations of safe operating procedures in their areas of responsibility should be informed that such action is not acceptable in a supervisor. On the other hand, a supervisor who is attuned to safety or who institutes innovative safety procedures should be given appropriate recognition, not only for his or her own sake, but as an example for other supervisors and nonsupervisory personnel.

Of course, one of the strongest motivators for safety (or any other aspect of performance) is the yearly performance review, whether or not it is tied in to a salary review. Safety performance and attitude should be made an important factor in this review, not just for supervisors, but for all personnel. Rewards or penalties for good or bad safety performance can be a powerful motivator.

MOTIVATING NONSUPERVISORY PERSONNEL

Among laboratory personnel will be found the entire gamut of attitudes toward safety. At one end of the spectrum are the hypochondriacs or chemophobes, who are so concerned about health or the hazards associated with chemicals that they are scarcely effective as laboratory workers. At the other end is the careless worker who feels that safety procedures represent a needless interference with the efficient performance of his or her job. Fortunately, the average laboratory employee falls into neither of these extreme categories, but wishes to work safely and may need to be shown how.

For the laboratory manager, the hypochondriac or chemophobe can present a major problem. If he or she is not taken seriously and does become the victim of an accident or work-related illness, the manager may be seen as culpable. On the other hand, if the employee's complaints are taken seriously, the manager can become embroiled in an endless series of arguments, tests of air and water, purchases of unneeded equipment, and so on. In severe cases, the only solution is to terminate the person's employment.

One way to prevent this type of problem is to screen such people out at the employment interview. A simple question about the state of the applicant's health will suffice. The hypochondriac will respond with a ten- to fifteen-minute discourse on his various problems or illnesses, while the average person will simply respond with "good" or "excellent," or perhaps a description of a chronic problem like high blood pressure. It is estimated that about 12 percent of the population suffers from hypochondriac syndrome. Undue fear of chemicals, at least, is probably not that common among chemists, since their education would tend to eliminate chemophobes before they get the bachelor's degree.

The person who willfully disregards safety in the laboratory on the grounds of expediency or efficiency can usually be handled with disciplinary action. He or she must be made to realize that safety is not a matter of personal choice, since management may be held liable for accidents or injuries, and that one careless employee endangers other personnel. A graded series of disciplinary steps may be taken—for example, an oral reprimand for the first offense, three days suspension without pay for the second, and termination for the third. Each such action should be carefully documented for protection against litigation, in case termination is necessary. In a unionized laboratory, of course, all such actions must be in accord with the contract. However, most union representatives these days are as concerned with safety as management is.

The task of motivating most employees—those who do not fall into either of the above classes—is mainly a matter of education, training, and "consciousness raising." Employees should be educated regarding sources of safety information, including Material Safety Data Sheets, available books and magazines, and the safety manual. They should be encouraged to bring any concerns they may have regarding the safety of materials or operations to the attention of their supervisor or the safety officer. Employees should be encouraged not to tolerate unsafe actions or procedures on the part of fellow employees, since their own health, lives, or livelihood may be endangered by such actions. Training of employees should be concerned with the proper use of safety equipment, conditions when it is required, and methods of recognizing safety hazards.

Raising the safety consciousness of employees, or teaching them to "think safety," can be done in many ways. Posters, prominently displayed, are one relatively simple method. Membership on the safety committee should be rotated among employees as a means of raising safety consciousness. Some organizations award prizes of various types to groups of employees who work a given number of person-hours without a lost-time injury. While this may be effective, it can have the disadvantage of encouraging employees to minimize the seriousness of injuries and not take adequate time for recovery.

Again, the most important factor in raising safety consciousness will be the attitude of supervisors and management. If it is apparent that the "bosses" regard safety as an important part of the job, employees will quickly get the message. The yearly performance review that accords safety a prominent position is a powerful tool.

8

Managing Hazardous Waste

Until about a decade ago the handling of hazardous waste was not a major concern of most laboratory managers. Waste chemicals, if water soluble, were simply poured down the nearest drain accompanied by copious quantities of tap water. Materials not soluble in water were thrown out with the trash, or, if liquid, allowed to evaporate in an open field, or were hauled away by a waste disposal firm whose disposal methods were not questioned.

However, those days are gone forever. The old maxim that "dilution is the solution to pollution" no longer applies. The rise of the environmental movement and the establishment of the EPA and various state environmental agencies with broad regulatory powers have forced organizations dealing with hazardous materials to pay attention to the generation and disposal of waste.

Most laboratories generate only small amounts of hazardous waste. The EPA has estimated that "small generators," that is, organizations that generate less than 1,000 kilograms of waste per month—a category into which most laboratories fall—are responsible for only about 0.4 percent of all hazardous waste. Nevertheless, this amounts to about 940,000 metric tons a year. Most laboratory waste consists of small quantities of hazardous material or acutely hazardous material that present special disposal problems.

Until September 1986 small generators enjoyed only minimal regulatory requirements under EPA. (Some states, however, did not recognize the small quantity exclusion.) In that month new regulations went into effect that applied to all generators of 100 kilograms or more per month of hazardous waste. Since 100 kilograms represents only ⅓ to ½ of a fifty-five-gallon drum, these regulations now encompass many chemical laboratories.

The new regulations require that the laboratory obtain an EPA identification number and have the wastes hauled and disposed of by EPA-approved contractors to EPA-approved sites. A Uniform Hazardous Waste Manifest must accompany each shipment of hazardous waste. The manifest must provide the name of the organization, a U.S. Department of Transportation description of the waste, the number and type of containers, and the weight. EPA regional headquarters should be contacted for an identification number and copies of the Uniform Hazardous Waste Manifest. State regulatory agencies should also be contacted for any necessary permits to store and dispose of hazardous waste.

In general, a waste is considered hazardous if it is ignitable, corrosive, toxic, or reactive, even if it is not listed in RCRA Appendix VIII (Federal Register 1980), Superfund list (Federal Register 1986b), or the list of Priority Pollutants (Clean Water Act 1978). Since these categories include virtually all chemicals used in laboratories, it can be assumed that all chemical waste generated in a laboratory is hazardous, at least by EPA definition. For details regarding definitions of hazardous waste, see the *Code of Federal Regulations* (*CFR*) (1986) (40 CFR part 261). The code contains a list of chemicals that are defined as hazardous, as well as EPA-approved methods of determining flammability, corrosivity, and toxicity. The EPA does not have an approved method for determining reactivity, and the methods for the other hazard parameters are still under development. The Federal Register should be consulted for changes in the test methods.

MINIMIZING HAZARDOUS WASTE

Although much of the waste generated by a laboratory is in small quantities, it will be found that disposal costs are very high, due to the fact that the waste is of varied chemical composition. The management of hazardous waste, therefore, begins with efforts to minimize the amount generated in the laboratory. This will require considerable attention on the part of laboratory managers and supervisors and the development of new habits of thought on the part of laboratory employees. The following are three methods of controlling the amount of hazardous waste that will need to be disposed of.

Inventory Control

Studies have shown that one of the major components of hazardous waste is simply unused chemicals that are not needed in the foreseeable future, or that are so old as to be of suspect quality. This situation arises because chemists have a tendency to overstock chemicals. Overstocking occurs for two reasons. First, chemical supply houses price chemicals so that the "large economy size" appears to be a good bargain. It may not be, if the cost of disposal of leftover material is taken into account. Second, chemists may overstock for fear that if the stock is used up, precious time may be lost before a new order is delivered.

Chemists must be trained to give accurate estimates of the quantities of chemicals needed over the short term, and to order only as much as is needed. Attention should also be paid to the chemical inventory so that timely reorders may be made. A documented or computer-controlled inventory system can be invaluable in controlling inventory. Supply houses should be evaluated on their promptness in filling orders as well as on quality and price. Chemists should be made aware of the costs of hazardous waste disposal to emphasize the need for minimizing the amount generated.

Minimizing Chemical Use

Research experiments and analytical methods should be examined with a view toward scaling down the amounts of chemicals used. This may require purchase of semimicro apparatus and equipment, but the savings in waste disposal costs should more than offset the expenditure.

Much of our traditional chemical laboratory work is done with quantities in the 100 gram/100 milliliter range. Most experiments can be scaled down by a factor of at least 10. Semimicro scale laboratory glassware is readily available for virtually all laboratory unit operations, especially organic chemical work. In analytical chemistry, sample preparation is often carried out on a macro scale, for the purpose of injecting a microliter or two into an instrument. There is no good reason why sample preparation cannot be carried out on at least a semimicro scale, as long as sample homogeneity is not a consideration.

One added advantage of scaling down laboratory operations is the improvement in safety. Hazards from fire, explosion, and employee exposure are all lessened. Many academic departments are adopting this scaled-down approach with gratifying results in safety, cost of chemicals, and cost of waste disposal. It was initially feared that students would lack the manipulative ability to perform the experiments correctly, but that fear proved to be unfounded.

Neutralizing Hazardous Waste

One other strategy for minimizing hazardous waste is to convert it to nonhazardous form for easier disposal or reuse.

One of the largest categories of laboratory hazardous waste is aqueous acids. These chemicals, especially commonly used acids like sulfuric, nitric, or hydrochloric, are easily neutralized with sodium bicarbonate and poured down the drain. However, care should be taken that hazardous heavy metals or other materials are not present. *Local waste water treatment authorities (publicly owned treatment works) should be contacted to ensure that local ordinances are not violated.* High concentrations of chromium from acid dichromate cleaning solutions, for example, may not be permitted to be drained. (Reasonable substitutes for dichromate cleaning solutions should be investigated.) Further treatment of the neutralized solution may be necessary if it is highly contaminated. Precipitation of a ferric iron or aluminum floc, for example, should be followed by filtration to trap heavy metals.

Aqueous alkali waste in most laboratories is not generated in nearly the quantities that acid is. This waste can be used to partially neutralize acid waste, provided the aqueous alkali does not contain sulfides, cyanides, or other materials that can give rise to hazardous gases on reaction with acid.

Especially hazardous materials (cyanides and sulfides, for example) can often be converted to relatively harmless compounds (cyanates and sulfates) by chemical reaction, especially if only small quantities are involved. See *Prudent Practices for Handling Hazardous Chemicals in Laboratories*, by the National Research Council (1981), for methods of neutralizing several of the common hazardous chemicals used in laboratories.

In many analytical laboratories, large quantities of organic solvents are used for extraction, separation, and dilution of samples prior to analysis. Usually only small quantities are consumed in the final determination, leaving large amounts of waste consisting mostly of solvent. In some laboratories these waste solvents are collected and purified by distillation for reuse. A highly efficient still must be available and care taken to produce a product that is acceptable for future use. Before going this route a careful analysis of the cost of this process versus the cost of disposal should be undertaken.

DISPOSAL OF HAZARDOUS WASTE

Despite all efforts to minimize hazardous waste at the source, many laboratories still find themselves in the small generator category, producing at least 100 kilograms per month and faced with the need to dispose of it. Not only is waste disposal costly; the laboratory may have difficulty finding a firm willing to handle small quantities of material. A further complication is created by EPA regulations, which forbid storage or accumulation of wastes for more than 180 days, or 270 days if the wastes need to be shipped more than 200 miles for disposal.

Many waste disposal firms require a flat fee, often as much as $500 plus administrative costs, regardless of the quantity involved. If lab-packs are used for the waste (see below), many waste disposal firms will insist on packing the drums themselves, which will add to the cost. Containers that are shipped must be accompanied by a manifest giving the name of the organization or laboratory, its EPA identification number, the intended disposal site, a U.S. Department of Transportation description of the waste, the number and type of containers, the weight of the waste, and the signature of a responsible member of the organization (USEPA 1986). (Although only one copy of the manifest is required, it is prudent to keep and file a copy.)

Two methods of disposing of waste material are generally available: incineration and landfilling (except for bulk liquids, which are not permitted to be landfilled). Of the two, incineration is to be preferred. When waste is incinerated, it is effectively destroyed and is no longer a liability of the generator. In landfills, however, the waste remains the responsibility of the generator, and the laboratory may find itself liable for future damage should the landfill fail and pollute groundwater.

It will be found advantageous to encourage laboratory workers to separate wastes into at least the following categories:

1. Organic solvents other than halogenated solvents
2. Halogenated organic solvents
3. Aqueous acids
4. Solid chemical wastes (kept in separate containers)
5. Exceptionally hazardous materials (including explosives and highly reactive chemicals such as peroxides, cyanides, and sulfides)

Aqueous acids may be neutralized and disposed of via the drains, if local authorities permit. If not, a waste disposal firm will have to be found that will accept this type of waste.

Solids (category 4 wastes) are most often disposed of via "lab-packs." These are 55-gallon drums, in which waste containers are packed in inert material like vermiculite or diatomaceous earth. The containers are packed so as to minimize the danger of breakage. The drums are buried in a landfill. They must only contain materials that are chemically compatible, and must not include unusually dangerous materials such as organic peroxides, reactive metals, or explosives. Bulk liquids are no longer permitted in lab-packs. For landfilling, EPA-specified steel drums must be used. For incineration, fiber drums are used.

Category 5 wastes may require innovative methods of disposal. Neutralization or conversion to a nonhazardous product, as described above, may be possible. Local police department bomb squads may be consulted for disposal of explosives.

Because of the requirement of identifying wastes on the manifest that accompanies the waste containers, it is important that the composition of a given waste container be known, at least by regulatory definition. This requirement can be met by keeping an accurate record of the various waste materials added to a mixture, including weights, or preferably by analysis of the waste prior to disposal. The latter is preferable because of the possibility of human error in recording wastes, but may require instrument capabilities the laboratory does not have. The use of an independent commercial laboratory for this analysis may be an acceptable alternative.

It is obvious that the control of hazardous waste is an important function of laboratory management. One strategy that should be explored is to locate other laboratories or organizations in a region that may be interested in a cooperative waste disposal arrangement. Since one of the major costs in waste disposal is transportation to the disposal site, a cooperative arrangement may prove economical. It may also be beneficial to exchange lists of chemicals that are to be disposed, since one laboratory's waste may be another laboratory's desired reagent. State regulatory agencies should be contacted regarding any needed permits and the legality of such activity.

9

Managing Safety in the Academic Laboratory

The academic laboratory is, in many respects, unique in comparison with industrial or other chemistry laboratories. Although the instructor may be analogous to the laboratory supervisor, the students differ considerably from their industrial counterparts. For example, the average student may spend only three to six hours per week in the laboratory, compared to the forty or more spent by workers in industrial laboratories. On the other hand, graduate students or teaching assistants may spend considerable time each week in the laboratory.

In addition, students vary in their chemical knowledge and skill, from a freshman with no high school chemistry background, to a senior who may be the equivalent of a laboratory professional.

Because of their teaching function, academic laboratories may have a much larger variety of hazardous chemicals and other hazards than the typical nonacademic laboratory. The variety of hazardous wastes produced is also usually greater in the academic laboratory.

Until recent decades, safety in the academic laboratory was largely a hit-or-miss affair. Depending on the knowledge and concern of the department personnel, practices vary widely from college to college, and even between courses in the same college. The risk of litigation has brought a greater consciousness of safety, however. Recent years have witnessed an increasing number of lawsuits against teachers and their employers on behalf of students who have been injured in laboratory accidents. It therefore behooves the modern college or university instructor to adhere to a set of safety regulations and enforce them rigorously in the laboratories under the instructor's control. Likewise, academic authorities should investigate and monitor safety in their institutions.

ORGANIZATION FOR SAFETY

The manner is which safety requirements are handled in academic institutions varies widely, generally depending on the size of the institution. In large universities, safety may be the responsibility of an institution-wide department with several employees reporting to the academic administration. In small liberal arts colleges, safety requirements may devolve upon the chemistry department. In many cases safety is a responsibility of the department head, either directly assigned or assumed by the individual holding that position. In other institutions, a safety officer may be appointed by the department head from among the faculty. In these cases, the position is usually held on a rotating basis, and the safety officer may or may not be assisted by a safety committee, also appointed on a rotating basis.

Regardless of the system in use, basic principles of safety management apply, just as they do in nonacademic laboratories. These can be summarized as follows:

- There should be one person with the prime responsibility for laboratory safety. Permitting individual faculty members to set their own safety rules is not conducive to a good, uniform safety program, nor to effective training of students in safe operations.
- The safety officer should be educated and trained in chemistry, and a member of the chemistry faculty or staff. (It is possible to appoint a nonfaculty person such as the stockroom manager to the position, but he or she should have a chemistry background.) If at all possible, the position of safety officer should be permanent, or an appointment of five years or more. The common practice of rotating the job on a yearly basis generally does not lead to as effective a program as one managed by a permanent appointee. Appointing the job is difficult if faculty members are used, since the position of safety officer is usually viewed as an onerous chore, and not important to one's career.
- The safety officer should have the backing of the department head and his or her superiors. As in nonacademic laboratories, the safety officer should be perceived as having the necessary authority to do the job properly.
- The use of a safety committee is highly recommended for the same reasons as in nonacademic laboratories.
- A safety manual should be produced for the guidance of laboratory instructors and students. In fact, it may be desirable to create separate manuals for instructors and students.

The booklet published by the American Chemical Society entitled "Safety in Academic Chemistry Laboratories" (1985) is highly recommended. Although the book is addressed to academic chemists, the latter half of the book is an excellent safety manual for students.

MANAGING STUDENT BEHAVIOR FOR SAFETY

Controlling the behavior of students in the laboratory can present special problems. Although many students behave responsibly, there are usually some who engage in horseplay in the laboratory or become careless, impatient, or forgetful. Such tendencies may lead to safety problems.

It is therefore doubly important that individual instructors who bear major responsibility for safe operations in their laboratories vigorously enforce safety rules. In additon to specific rules for a given laboratory or experiment, a set of general rules for student behavior should be enforced. General rules should include, but are not necessarily limited to, the following:

- Appropriate eye protection should be worn at all times in the laboratory. This should apply to all students, instructors, and visitors. Contact lenses, unless covered by goggles that protect against gases and vapors, should not be permitted.
- No smoking or preparation, storage, or consumption of food or drink should be permitted in the laboratory.
- No student should be permitted to work alone.
- No "unauthorized" experiments should be allowed. All laboratory work should be approved by the instructor.
- Horseplay, roughhousing, or other "fun and games" should never be allowed in the laboratory. This should include the playing of radios and tape recorders, and the use of earphones, which may interfere with communication.
- Laboratory glassware should never be used as containers for food or drink (in or out of the laboratory). Chipped or broken glassware should never be used without fire polishing.
- Appropriate clothing should be worn. The use of laboratory coats or aprons should be encouraged. Shorts or other clothing that leaves substantial areas of skin exposed should not be allowed. Shoes or sandals that do not cover the toes should be prohibited, as should canvas or other fabric shoes.
- Loose hair should be pinned back or confined. Jewelry that fits tight to the skin (rings and bracelets) should be removed.

- Coats, jackets, or raincoats should be hung outside of the laboratory. Backpacks and briefcases should be kept out of the laboratory.
- The mouth is never to be used as a source of suction for pipetting, siphoning, or any other purpose.

These are general rules that need to be enforced in all laboratories. Laboratories engaged in radiochemistry, X-ray diffraction, or laser experimentation may require additional procedures specific to those applications.

Instructors also must be continually on the alert for infractions of the rules. Students will be careless or willfully disregard safety rules in the interest of speed or convenience, and such violations should be dealt with immediately. Instructors should bear in mind that they may be found legally responsible for accidents or injuries in a laboratory under their control.

Instructors must also be on the alert for unsafe conditions such as the following: malfunctioning electrical equipment; blocked access to fire extinguishers, eyewash fountains, or exits; use of chipped or broken glassware; and unlabeled chemical containers. Hazardous conditions should be corrected immediately. It is good practice for the instructor to conduct a ten- to fifteen-minute safety inspection during each laboratory period. Instructors of course should be careful to obey all safety regulations themselves. It is also good practice to post a list of applicable safety regulations in a conspicuous place in the laboratory.

There is a growing trend in teaching circles to limit quantities of chemicals used for experiments in teaching laboratories. This not only minimizes the hazardous waste problems of the laboratory, but also reduces the extent of damage or injury resulting from accidents. Most chemical principles do not depend on the absolute quantity reacting and may be demonstrated on a semimicro scale rather than with the larger quantities usually used.

EDUCATING FOR SAFETY

In addition to ensuring safety in academic laboratories, the academic community should bear responsibility for teaching the proper attitude toward safety in all chemical work. Such an attitude is an important factor in chemical work, and the graduate who does not possess it may be perceived as lacking professional training. In turn, this lack of training will reflect adversely on the institution that trained the chemist.

In some institutions a course in chemical safety is offered. In addition to shaping students' attitudes toward safety, a course of this kind may have considerable educational value, since aspects of toxicology, biochemistry, biophysics, economics, and politics may be included as legitimate topics.

Additional courses are not easily fitted into the chemistry program in most institutions because of stringent curricular demands and limited time. It is therefore usually necessary to integrate the teaching of safety into existing courses, and any laboratory course can serve this purpose. The following are some techniques for incorporating safety into course work.

- Most important is that the teacher and laboratory assistants set a good example. This involves following all of the rules and being alert for safety infractions.
- All written experimental procedures should include a section on safety. Even if no special safety precautions are necessary, this should be explicitly stated.
- Students should be exposed to the literature on safety at some point in the undergraduate curriculum, probably at the start of the junior year, when students begin to work more independently in the laboratory. A collection of books on safety should be a part of the chemistry department library and should be made available to the students. Material Safety Data Sheets on all chemicals used in the laboratory also should be available.
- Students should be required to purchase a book such as *Prudent Practices for Handling Hazardous Chemicals in Laboratories* (National Research Council 1981) or the ACS *Safety in Academic Chemistry Laboratories* (1985), and given a series of reading assignments and quizzes on safety as part of the course requirements.
- Students should be asked to prepare a Material Safety Data Sheet for a fictitious mixture of chemicals assigned by the instructor. This will require considerable exercise of technical knowledge as well as the ability to read and understand an MSDS.
- All experiments and projects proposed by students should have a section on safety, which should be carefully scrutinized by instructors before approval. Special attention should be paid to proposals to deal with any toxic wastes that may be produced. Engineering students concerned with design of chemical processes should be aware of any problems connected with hazardous waste.

There are many opportunities in both classroom and laboratory for bringing safety to the attention of students and developing the awareness of safety and environmental protection as a necessary part of the professional chemist's and chemical engineer's job.

10

Laboratory Design and Safety

With the decision to remodel, add a wing, or build a new laboratory structure, the laboratory manager embarks on an extremely complex and problematical venture. The goal is an environment that includes the following features: suitable space for both bench work and offices; ample utilities for the most exacting work; storage space in which supplies, laboratory records, and hazardous materials are secure; and properly installed, effective safety features to protect property.

In undertaking such a project, the manager needs to recognize that for the building professions and trades, the special requirements of a laboratory represent a complex and demanding set of problems unlike those posed by standard types of construction. The laboratory manager must guide the contract or corporate architects by preparing an exacting set of criteria or basic data (not to be confused with specifications, which are the domain of building specialists). Required by laboratory designers as a guide to satisfying user needs and environmental requirements, basic data are the sole responsibility of the laboratory manager or laboratory expansion management team.

The rationale for building will be found in the business plan of the organization. Expansion may be needed because the pressures of additional laboratory personnel and equipment have resulted in declining efficiency. For some laboratories, desired research contracts require advanced technical equipment and special operating conditions not available in the existing facilities. Or a move into a new technology may require expansion. Once the decision has been made, laboratory management has the responsibility to incorporate safety into plans for the new facilities.

OBJECTIVES FOR LABORATORY SAFETY PLANNING

The following is a list of objectives to be accomplished in the design stage for safety in a new facility.

- Develop a workable safety statement
- Define the technical objective(s) of the new laboratory
- Identify techniques for determining user safety needs
- Rank safety requirements for the laboratory facility
- Systematically organize a safety planning program
- Evaluate existing facility safety performance
- Explore critical safety requirements
- Incorporate technological safety advances into the basic safety data

Developing a Workable Safety Statement

In a well-managed laboratory the chairman, president, or another member of top management will have on public record a statement of safety philosophy. It may be a few sentences, or it may run to several pages in length and express a deep commitment to safety. The statement may be taken from the safety manual. Whatever the source, it should be complete and clear, so that the organization's position on safety is understood by everyone concerned. It may be concise, but not so short and simplistic that it is ignored. An example of a short statement is: "Safety requires focusing attention on the immediate work environment and modifying behavior and facilities to protect personnel and property." All recommendations, requirements, and suggestions to the laboratory expansion management team should be based on such a statement, one incorporating safety principles.

Defining the Technical Objectives of the New Laboratory

The choice of safety installations, facilities, and equipment is dictated by the laboratory objectives, which may be one or more of the following.

General-purpose laboratories are those in which a variety of operations are carried out, usually in conventional apparatus and glassware, employing a number of the usual small laboratory instruments and using relatively small amounts of chemicals.

Special-purpose laboratories are those intended for continuing use in one operation or manner involving a definite and limited number of specific hazards. These may be low-degree hazards such as botanical growth or instrumentation, which might require less stringent fire protection, electrical, or emergency features.

Special occupancy laboratories may involve high-level hazards such as high-pressure equipment, carcinogens or radioactive substances, flammable liquids or gases, combustible dusts, high-energy materials, high-voltage equipment, or biological health hazards.

"One-of-a-kind" laboratories are those that are technically adequate for advancing the state of the art as a test facility, or a prototype laboratory for rapidly advancing technologies such as electronics or laser, nuclear, or medical systems. In most instances the shape of the unit is determined by the size of the equipment, flow of materials, and personnel safety.

Identifying Techniques for Determining User Safety Needs

User needs for laboratory safety have been discussed in numerous technical journals, general science journals, engineering journals, and in journals of interest to architects, civil engineers, safety engineers, utility engineers, and educators. Computer data bases are a useful tool for tapping into this literature.

Engineering associations and similar organizations have issued invaluable literature on laboratory safety concepts. Two examples, among many, are the National Safety Council (Chicago) and the National Fire Protection Association (Boston). Federal (OSHA), state, and local regulations are not considered here, since these are the concern of the architect. However, the manager should be aware of them.

The users themselves are another source of information concerning user safety needs. Laboratory personnel who have previous experience in a similar situation should be formed into a subcommittee to share information with management. If in-house user experience is lacking, a survey of new and similar laboratories that are not in direct competition may locate appropriate user situations that can be investigated through correspondence and tours. For specific pieces of equipment or instrumentation, users can be found who will discuss safety needs.

When a laboratory is expanding to increase physical space, it is all too easy to overlook the experience of bench chemists or technicians, since most of these individuals have survived the rigors of laboratory work without serious harm. But they can play an important part in the basic safety data program. A job safety analysis can be done on each complete operation or a group of similar operations. An observer should note each step of the operation from start to finish without comment. The analysis should be evaluated by the observer and at least one other person in terms of ergonomics (reaching, standing, carrying, and so forth); utility location and convenience; lighting, ventilation, and available space; fire and explosion containment; personal safety equipment requirements; and location of chemicals, reagents, small equipment, and other laboratory items. This procedure will reveal patterns not obvious before. The results can be included in the basic safety data.

Ranking Safety Requirements for Laboratory Facilities

In the view of many laboratory managers, the heating, ventilation, and air-conditioning system (HVAC) is the most important requirement for laboratory safety. It is this system that supplies the laboratory with sufficient input air to (1) permit proper movement of air, fumes, and gases through hood exhausts, (2) prevent migration of noxious and hazardous vapors into other rooms, and (3) provide environmental comfort for personnel.

Hoods are central to the basic safety of any laboratory. Their complexity and use increases with hazard severity. Since air pollution is a possibility and dilution is no longer an acceptable treatment method, hood exhaust from moderate to high hazard areas may require pretreatment. The pretreatment may be filtration, a scrubber facility, chemical treatment, heat treatment, or a combination of these. Costs will obviously be high for installation, maintenance, and operation of any pretreatment system. The important point in assembling basic safety data is to present all the facts and the rationale behind recommendations without hedging.

Another important safety requirement concerns fire protection systems. These must be carefully designed for laboratory facilities because of the high fire potential and the cost of replacing expensive equipment and instruments.

A fire protection requirement not usually given adequate consideration is storage of laboratory records (archives), including quality assurance/quality control data, computer printouts, and logs of all kinds required for patents, legal protection, and demonstration of compliance with federal analytical protocols.

A crucial part of storage fire protection concerns areas where chemicals, reagents, solvents, compressed gases, and hazardous wastes are retained. Special attention must be given to types of extinguishers and backup systems. Are fire wall needs more restrictive than code? Are these areas located away from normal pedestrian traffic or even remote from the laboratory? Whatever the degree of hazard, the potential must be stated in the basic safety data.

For personnel protection in case of fire, thought should be given to the warning communication system. Some means of sounding an alarm must be provided, and it is vital that the alarm be heard in all areas of the facility, including rest rooms, walk-in refrigerators, and the like. Thought should also be given to a backup, nonelectrical system such as hand- or compressed gas-operated sirens for use when electrical power is disabled.

Following these items, ranking of safety requirements is dependent on the level of hazard encountered in the type of work to be carried out in the facility. Radiochemistry and nuclear and biological health systems will be high on the list. In analytical laboratories dealing with environmentally hazardous materials on a small scale, personal protective gear and employee awareness may be more important than intensive engineering design. Hoods are the main protective device in these laboratories. However, there are always exceptions. High-energy analytical systems requiring shielding may be in use. Cryogenic liquids may be needed for special instrumentation. Flammable or explosive gases require special handling facilities. All of these should be noted on the basic safety data.

General purpose laboratories in the main are covered by federal, state, and local structure, sanitation, occupancy, and fire and electrical codes or regulations. These need not be included in the basic safety data other than to state that they must be observed.

At a minimum, the ranked list of laboratory basic safety data should include the following items:

- *Laboratory exhaust ventilation:* Hoods, exhaust considerations, air supply, and air discharge
- *Fire safety:* Protective equipment, alarms, maintenance, and testing
- *Emergency response:* Spill control, breathing apparatus, protective clothing, and cleanup procedures
- *Chemical handling:* Solvent storage, chemical storage, waste retention, waste disposal; special safety procedures for compressed liquids and gases, acutely hazardous materials, and carcinogens
- *Special laboratory situations:* High pressure, high voltage, high energy, shielding, containment, lighting, biohazards, animal rooms, decontamination areas, change rooms, and washing areas

The laboratory manager can rank the above by knowledge of user needs and his or her experience and professional judgement.

Systematically Organizing a Safety Planning Program

A laboratory expansion management team will usually have three or more members selected by the laboratory head. The members may be engineers, architects, chief scientists, support managers, human relations specialists, or consultants. The team will have a target date for a presentation of basic data to management. From the target date a planning and design program will be developed that will allow the team to meet the commitment. The program will also flag areas where extra help and special assistance will be needed.

The basic safety data must be coordinated with the data from other members of the team and must be available prior to the indicated dates shown in the overall planning and design program. Even though the program may not appear logical with regard to safety, the laboratory manager responsible for basic safety data must accept the program, thoroughly understand the sequence of events, and plan his efforts to meet schedules.

Example of a Planning and Design Program Outline

1. Site selection
   - Wind velocity and direction—effect on exhaust dispersal
   - Subsoil—effect of toxic liquid spills
   - Electromagnetic interference—effect on instrumentation
   - Security—storage and archives
   - Hazards—proximity and mitigation of blast, fire, radiation, and exposure to toxic substances and pathogens
2. Structure geometry
   - Interior arrangement—relation between low- and high-hazard areas, load bearing (shielding), common or unitized emergency systems, and instruments
   - Allowance for expansion—(average time period: every five years)
   - Versatility—effect of rearrangements of working areas on existing safety systems
   - Convertibility—effect of shifting equipment on safety systems
   - Modularity—interiors that can safely be rearranged without the assistance of the building trades
3. Functional relationships
   - Administrative offices location

- Services (food, mail, maintenance, and so forth)—isolated from hazardous areas, independent air supply, or positive pressure air in relation to work areas
- Interrelationships—research, pilot plant, process development, and product development
- Shared operations—physical testing, engineering, high-hazard areas (personnel safety training among different departments)
- Support services—glassware washing (decontamination procedures, toxicity awareness, corrosive cleaning agents, solvent precleaning, and waste wash disposal)
- Sterilizing—disinfectant disposal, pressure (steam), drain monitors, and disposal system
- Storage—fire protection, grounding system, ventilation, federal requirements, compressed gases, oxidizers, and archive protection
- Maintenance—laboratory safety training and instrument manufacturers' training

4. Module definition
- Unit laboratory—number of workers, code or regulation requirements
- Work surfaces—heights to be specified
- Work clearance—extra width doors and maintenance spacing
- Hoods—need and placement
- Storage space
- Laboratory offices—enclosed, semi-enclosed, or away from laboratory
- Administrative offices—separate air supply, fire walls, and safety and emergency communications

Evaluating Existing Facility Safety Performance

Injury reports may be scrutinized for evaluation of safety performance. However, review of lost-time injury records will generally be of little help since the number of reports will probably not be statistically significant. Minor injury reports will be more numerous, but review will probably not show any clear trends. The ideal report is one that evaluates the potential for serious injury using an arbitrary scale defined by the laboratory. Since such reports are rarely available, the basic safety data evaluator must conduct an audit of the existing facilities to estimate current safety performance.

The safety audit suggested here focuses on in-place safety concepts, features, and devices. How effective are they? For example, how do users feel about available fire protection? Are users comfortable with deluge extinguishers over their work areas? How would their work area be affected if a sprinkler head was activated? What alternative would the user suggest? What about hazardous wastes? Are these areas managed in such a way that wastes go directly to the handling areas, or are the wastes disposed of in some other manner?

The recommended audit does not involve checking safety book operating procedures and regulations for compliance. Rather, it is a study of the physical plant for good and bad points, for possible improvements, and for additions that could bring the laboratory up to the most recently established technical safety standards. Such an audit will bring up many questions, and assistance may be required from trained safety personnel.

To perform the audit, the basic safety member of the expansion team should enlist the aid of an engineer who is thoroughly familiar with the laboratory building and a laboratory first-line supervisor to form a team to review the building in accordance with a plan such as that shown below. The team should study each item listed before moving on to the next.

Sample Design Safety Audit

1. Chemical hazards
   - exotherms
   - flammables
   - peroxidizables
   - toxics
   - corrosives
   - carcinogens
   - waste disposal
2. Personal protective equipment
   - eyes and face
   - ears
   - head
   - hands and arms
   - feet and legs
   - respiratory system
   - trunk
3. Safety equipment adequacy
   - fire extinguishers
   - safety showers
   - blankets and absorbent materials
   - air masks
   - emergency lights

fire doors
spill pillows

4. Procedures
   specific equipment procedures
   lock and tag procedures
   labels
   noncompany personnel safety
   safety awareness
5. Physical hazards
   storage
   glassware safety
   congestion
   exposed or unlabeled hot surfaces
6. Process and lab equipment
   hoods
   pressure vessels
   compressed gas cylinders
   electrical devices
   equipment guards
   ovens
   step ladders
   tools
   shields, secondary containers

Upon completion of the report, the team can summarize its findings, and new information may be added to the basic safety data.

Explore Critical Safety Requirements

In evaluating safety requirements it is important to get a second opinion, preferably from an experienced colleague. In technical matters we are specialists, and when we go outside of our field we seek expert guidance. The same attitude should apply to safety. Specialists can help with alternatives, entirely new concepts, cost comparisons, and so on. All possible approaches for critical areas should be reviewed to find how others' solutions to a similar problem may apply.

Incorporating Technological Safety Advances into the Basic Safety Data

All study to this point has been directed toward ascertaining the level of safety in the existing laboratory. Study of safety literature, visits to outside laboratories, discussions with laboratory users, and other efforts have given leads on possible improvements in safety features. In following up these leads, the team member responsible for basic safety should consult the various suppliers of laboratory safety equipment to find out about up-to-date safety technology specific to the laboratory's needs. A day with a safety consultant could be the most rewarding day spent on the project.

The expanding information on the toxicity of chemicals and materials has made laboratories more aware of exposure hazards. There are many ways this exposure can be minimized. High on the list of concerns is air movement and exhaust. Ventilation engineers have solved many problems of air movement for personnel comfort. They have been able to zone systems to minimize mixing of clean and contaminated air and have improved the exhaust of air from work areas.

Engineers have devised many special systems to capture noxious or toxic vapors at the source. Such systems can move air and its contaminants away from areas where it is not practical to have a laboratory hood. Spot, or localized, ventilation can be designed into an overall laboratory exhaust system, a more effective solution than resorting to jerry-rigged laboratory equipment to handle individual sources of toxic fumes.

An increasingly favored approach is the separate exhaust system, complete with its own fans, ducts, fire dampers, and other controls similar to the hood system, but completely independent. The system may be ducted to all workbenches and is tapped into by standard connectors and flexible reinforced tubing. The tubing is available in diameters to meet the needs of most situations. One novel arrangement uses rectangular ducts as reagent shelves with ports on the vertical faces. Another terminates the ducts in a pier between four movable benches. The flexible tubing can be run in the center of the benches to keep it from interfering with the work.

Hoods have undergone many design improvements to increase exhaust efficiency and maintain ease of operation. One new option to be considered is the low-level flow alarm, which is set at an acceptable flow level for exhaust protection. If the air flow falls below the set point, an alarm is sounded to alert the user to the need for action.

There have been recent advances in many other areas of interest to laboratory managers. Material handling, movement, and storage have all seen progress. Waste disposal regulations by all levels of government have forced improvements in that sphere. Interior finishes, flooring, and casework surfaces have improved in solvent resistance and fire resistance. It is strongly recommended that time be spent with suppliers to evaluate the applicability of these and other developments. The rewards could be great.

Appendix 1

OSHA 29 CFR 1910.1200 HAZARD COMMUNICATION STANDARD

(a) Purpose

(1) The purpose of this section is to ensure that the hazards of all chemicals produced or imported by chemical manufacturers or importers are evaluated, and that information concerning their hazards is transmitted to affected employers and employees within the manufacturing sector. This transmittal of information is to be accomplished by means of comprehensive hazard communication programs, which are to include container labeling and other forms of warning, material safety data sheets and employee training.

(2) This occupational safety and health standard is intended to address comprehensively the issue of evaluating and communicating chemical hazards to employees in the manufacturing sector, and to preempt any state law pertaining to this subject. Any state which desires to assume responsibility in this area may only do so under the provisions of § 18 of the Occupational Safety and Health Act (29 U.S.C. 651 et. seq.) which deals with state jurisdiction and state plans.

(b) Scope and Application

(1) This section requires chemical manufacturers or importers to assess the hazards of chemicals which they produce or import, and all employers in SIC Codes 20 through 39 (Division D, Standard Industrial Classification Manual) to provide information to their employees about the hazardous chemicals to which they are exposed, by means of a hazard communication program, labels and other forms of warning, material safety data sheets, and information and training. In addition, this section requires distributors to transmit the required information to employers in SIC Codes 20–39.

(2) This section applies to any chemical which is known to be present in the workplace in such a manner that employees may be exposed under normal conditions of use or in a foreseeable emergency.

(3) This section applies to laboratories only as follows:

(i) Employers shall ensure that labels on incoming containers of hazardous chemicals are not removed or defaced;

(ii) Employers shall maintain any material safety data sheets that are received with incoming shipments of hazardous chemicals, and ensure that they are readily accessible to laboratory employees; and,

(iii) Employers shall ensure that laboratory employees are apprised of the hazards of the chemicals in their workplaces in accordance with paragraph (h) of this section.

(4) This section does not require labeling of the following chemicals:

(i) Any pesticide as such term is defined in the Federal Insecticide, Fungicide, and Rodenticide Act (7 U.S.C. 136 et seq.), when subject to the labeling requirements of that Act and labeling regulations issued under that Act by the Environmental Protection Agency;

(ii) Any food, food additive, color additive, drug, or cosmetic, including materials intended for use as ingredients in such products (e.g., flavors and fragrances), as such terms are defined in the Federal Food, Drug, and Cosmetic Act (21 U.S.C. 301 et seq.) and regulations issued under that Act, when they are subject to the labeling requirements of that Act and labeling regulations issued under that Act by the Food and Drug Administration.

(iii) Any distilled spirits (beverage alcohols), wine, or malt beverage intended for nonindustrial use, as such terms are defined in the Federal Alcohol Administration Act (27 U.S.C. 201 et seq.) and regulations issued under that Act, when subject to the labeling requirements of that Act and labeling regulations issued under that Act by the Bureau of Alcohol, Tobacco, and Firearms; and,

(iv) Any consumer product or hazardous substance as those terms are defined in the Consumer Product Safety Act (15 U.S.C. 2051 et seq.) and Federal Hazardous Substances Act (15 U.S.C. 1261 et seq.) respectively, when subject to a consumer product safety standard or labeling requirement of those Acts, or regulations issued under those Acts by the Consumer Product Safety Commission.

(5) This section does not apply to:

(i) Any hazardous waste as such term is defined by the Solid Waste Disposal Act, as amended by the Resource Conservation and Recovery Act of 1976, as amended (42 U.S.C. 6901 et seq.), when subject to regulations issued under that Act by the Environmental Protection Agency;

(ii) Tobacco or tobacco products;

(iii) Wood or wood products;

(iv) Articles; and,

(v) Foods, drugs, or cosmetics intended for personal consumption by employees while in the workplace.

(c) Definitions

"Article" means a manufactured item: (i) Which is formed to a specific shape or design during manufacture; (ii) which has end use function(s) dependent in whole or in part upon its shape or design during end use; and (iii) which does not release, or otherwise result in exposure to, a hazardous chemical under normal conditions of use.

"Assistant Secretary" means the Assistant Secretary of Labor for Occupational Safety and Health, U.S. Department of Labor, or designee.

"Chemical" means any element, chemical compound or mixture of elements and/or compounds.

"Chemical manufacturer" means an employer in SIC Codes 20 through 39 with a workplace where chemical(s) are produced for use or distribution.

"Chemical name" means the scientific designation of a chemical in accordance with the nomenclature system developed by the International Union of Pure and Applied Chemistry (IUPAC) or the Chemical Abstracts Service (CAS) rules of nomenclature, or a name which will clearly identify the chemical for the purpose of conducting a hazard evaluation.

"Combustible liquid" means any liquid having a flashpoint at or above 100°F (37.8°C), but below 200°F (93.3°C), except any mixture having components with flashpoints of 200°F (93.3°C), or higher, the total volume of which make up 99 percent or more of the total volume of the mixture.

"Common name" means any designation or identification such as code name, code number, trade name, brand name or generic name used to identify a chemical other than by its chemical name.

"Compressed gas" means:

(i) A gas or mixture of gases having, in a container, an absolute pressure exceeding 40 psi at 70°F (21.1°C); or

(ii) A gas or mixture of gases having, in a container, an absolute pressure exceeding 104 psi at 130°F (54.4°C) regardless of the pressure at 70°F (21.1°C); or

(iii) A liquid having a vapor pressure exceeding 40 psi at 100°F (37.8°C) as determined by ASTM D-323-72.

"Container" means any bag, barrel, bottle, box, can, cylinder, drum, reaction vessel, storage tank, or the like that contains a hazardous chemical. For purposes of this section, pipes or piping systems are not considered to be containers.

"Designated representative" means any individual or organization to whom an employee gives written authorization to exercise such employee's rights under this section. A recognized or certified collective bargaining agent shall be treated automatically as a designated representative without regard to written employee authorization.

"Director" means the Director, National Institute for Occupational Safety and Health, U.S. Department of Health and Human Services, or designee.

"Distributor" means a business, other than a chemical manufacturer or importer, which supplies hazardous chemicals to other distributors or to manufacturing purchasers.

"Employee" means a worker employed by an employer in a workplace in SIC Codes 20 through 39 who may be exposed to hazardous chemicals under normal operating conditions or foreseeable emergencies, including, but not limited to production workers, line supervisors, and repair or maintenance personnel. Office workers, grounds maintenance personnel, security personnel or non-resident management are generally not included, unless their job performance routinely involves potential exposure to hazardous chemicals.

"Employer" means a person engaged in a business within SIC Codes 20 through 39 where chemicals are either used, or are produced for use or distribution.

"Explosive" means a chemical that causes a sudden, almost instantaneous release of pressure, gas, and heat when subjected to sudden shock, pressure, or high temperature.

"Exposure" or "exposed" means that an employee is subjected to a hazardous chemical in the course of employment through any route of entry (inhalation, ingestion, skin contact or absorption, etc.), and includes potential (e.g., accidental or possible) exposure.

"Flammable" means a chemical that falls into one of the following categories:

(i) "Aerosol, flammable" means an aerosol that, when tested by the method described in 16 CFR 1500.45, yields a flame projection exceeding 18 inches at full valve opening, or a flashback (a flame extending back to the valve) at any degree of valve opening:

(ii) "Gas, flammable" means:

(A) A gas that, at ambient temperature and pressure, forms a flammable mixture with air at a concentration of thirteen (13) percent by volume or less; or

(B) A gas that, at ambient temperature and pressure, forms a range of flammable mixtures with air wider than twelve (12) percent by volume, regardless of the lower limit;

(iii) "Liquid, flammable" means any liquid having a flashpoint below 100°F (37.8°C), except any mixture having components with flashpoints of 100°F (37.8°C) or higher, the total of which make up 99 percent or more of the total volume of the mixture.

(iv) "Solid, flammable" means a solid, other than a blasting agent or explosive as defined in § 1910.109(a), that is liable to cause fire through friction, absorption of moisture, spontaneous chemical change, or retained heat from manufacturing or processing, or which can be ignited readily and when ignited burns so vigorously and persistently as to create a serious hazard. A chemical shall be considered to be a flammable solid if, when tested by the method described in 16 CFR 1500.44 it ignites and burns with a self-sustained flame at a rate greater than one-tenth of an inch per second along its major axis.

"Flashpoint" means the minimum temperature at which a liquid gives off a vapor in sufficient concentration to ignite when tested as follows:

(i) Tagliabue Closed Tester (see American National Standard Method of Test for Flash Point by Tag Closed Tester, Z11.24–1979 (ASTM D 56–79)) for liquids with a viscosity of less than 45 Saybolt Universal Seconds (SUS) at 100°F (37.8°C), that do not contain suspended solids and do not have a tendency to form a surface film under test; or

(ii) Pensky-Martens Closed Tester (see American National Standard Method of Test for Flash Point by Pensky-Martens Closed Tester, Z11.7–1979 (ASTM D 93–79)) for liquids with a viscosity equal to or greater than 45 SUS at 100°F (37.8°C), or that contain suspended solids, or that have a tendency to form a surface film under test; or

(iii) Setaflash Closed Tester (see American National Standard Method of Test for Flash Point by Setaflash Closed Tester (ASTM D 3278–78)).

Organic peroxides, which undergo autoaccelerating thermal decomposition, are excluded from any of the flashpoint determination methods specified above.

"Foreseeable emergency" means any potential occurrence such as, but not limited to, equipment failure, rupture of containers, or failure of control equipment which could result in an uncontrolled release of a hazardous chemical into the workplace.

"Hazard warning" means any words, pictures, symbols, or combination thereof appearing on a label or other appropriate form of warning which convey the hazards of the chemical(s) in the container(s).

"Hazardous chemical" means any chemical which is a physical hazard or a health hazard.

"Health hazard" means a chemical for which there is statistically significant evidence based on at least one study conducted in accordance with established scientific principles that acute or chronic health effects may occur in exposed employees. The term "health hazard" includes chemicals which are carcinogens, toxic or highly toxic agents, reproductive toxins, irritants, corrosives, sensitizers, hepatotoxins, nephrotoxins, neurotoxins, agents which act on the hematopoietic system, and agents which damage the lungs, skin, eyes, or mucous membranes. Appendix A provides further definitions and explanations of the scope of health hazards covered by this section, and Appendix B describes the criteria to be used to determine whether or not a chemical is to be considered hazardous for purposes of this standard.

"Identity" means any chemical or common name which is indicated on the material safety data sheet (MSDS) for the chemical. The identity used shall permit cross-references to be made among the required list of hazardous chemicals, the label and the MSDS.

"Immediate use" means that the hazardous chemical will be under the control of and used only by the person who transfers it from a labeled container and only within the work shift in which it is tranferred.

"Importer" means the first business with employees within the Customs Territory of the United States which receives hazardous chemicals produced in other countries for the purpose of supplying them to distributors or manufacturing purchasers within the United States.

"Label" means any written, printed, or graphic material displayed on or affixed to containers of hazardous chemicals.

"Manufacturing purchaser" means an employer with a workplace classified in SIC Codes 20 through 39 who purchases a hazardous chemical for use within that workplace.

"Material safety data sheet (MSDS)" means written or printed material concerning a hazardous chemical which is prepared in accordance with paragraph (g) of this section.

"Mixture" means any combination of two or more chemicals if the combination is not, in whole or in part, the result of a chemical reaction.

"Organic peroxide" means an organic compound that contains the bivalent -O-O-structure and which may be considered to be a structural derivative of hydrogen peroxide where one or both of the hydrogen atoms has been replaced by an organic radical.

"Oxidizer" means a chemical other than a blasting agent or explosive as defined in § 1910.109(a), that initiates or promotes combustion in other materials, thereby causing fire either of itself or through the release of oxygen or other gases.

"Physical hazard" means a chemical for which there is scientifically valid evidence that it is a combustible liquid, a compressed gas, explosive, flammable, an organic peroxide, an oxidizer, pyrophoric, unstable (reactive) or water-reactive.

"Produce" means to manufacture, process, formulate, or repackage.

"Pyrophoric" means a chemical that will ignite spontaneously in air at a temperature of 130°F (54.4°C) or below.

"Responsible party" means someone who can provide additional information on the hazardous chemical and appropriate emergency procedures, if necessary.

"Specific chemical identity" means the chemical name, Chemical Abstracts Service (CAS) Registry Number, or any other information that reveals the precise chemical designation of the substance.

"Trade secret" means any confidential formula, pattern, process, device, information or compilation of information (including chemical name or other unique chemical identifier) that is used in an employer's business, and that gives the employer an opportunity to obtain an advantage over competitors who do not know or use it.

"Unstable (reactive)" means a chemical which in the pure state, or as produced or transported, will vigorously polymerize, decompose, condense, or will become self-reactive under conditions of shock, pressure or temperature.

"Use" means to package, handle, react, or transfer.

"Water-reactive" means a chemical that reacts with water to release a gas that is either flammable or presents a health hazard.

"Work area" means a room or defined space in a workplace where hazardous chemicals are produced or used, and where employees are present.

"Workplace" means an establishment at one geographical location containing one or more work areas.

(d) Hazard Determination

(1) Chemical manufacturers and importers shall evaluate chemicals produced in their workplaces or imported by them to determine if they are hazardous. Employers are not required to evaluate chemicals unless they choose not to rely on the evaluation performed by the chemical manufacturer or importer for the chemical to satisfy this requirement.

(2) Chemical manufacturers, importers or employers evaluating chemicals shall identify and consider the available scientific evidence concerning such hazards. For health hazards, evidence which is statistically significant and which is based on at least one positive study con-

ducted in accordance with established scientific principles is considered to be sufficient to establish a hazardous effect if the results of the study meet the definitions of health hazards in this section. Appendix A shall be consulted for the scope of health hazards covered, and Appendix B shall be consulted for the criteria to be followed with respect to the completeness of the evaluation, and the data to be reported.

(3) The chemical manufacturer, importer or employer evaluating chemicals shall treat the following sources as establishing that the chemicals listed in them are hazardous:

(i) 29 CFR Part 1910, Subpart Z, Toxic and Hazardous Substances, Occupational Safety and Health Administration (OSHA); or,

(ii) *Threshold Limit Values for Chemical Substances and Physical Agents in the Work Environment*, American Conference of Governmental Industrial Hygienists (ACGIH) (latest edition).

The chemical manufacturer, importer, or employer is still responsible for evaluating the hazards associated with the chemicals in these source lists in accordance with the requirements of the standard.

(4) Chemical manufacturers, importers and employers evaluating chemicals shall treat the following sources as establishing that a chemical is a carcinogen or potential carcinogen for hazard communication purposes:

(i) National Toxicology Program (NTP), *Annual Report on Carcinogens* (latest edition);

(ii) International Agency for Research on Cancer (IARC) *Monographs* (latest editions); or

(iii) 29 CFR Part 1910, Subpart Z, Toxic and Hazardous Substances, Occupational Safety and Health Administration.

Note.—The *Registry of Toxic Effects of Chemical Substances* published by the National Institute for Occupational Safety and Health indicates whether a chemical has been found by NTP or IARC to be a potential carcinogen.

(5) The chemical manufacturer, importer or employer shall determine the hazards of mixtures of chemicals as follows:

(i) If a mixture has been tested as a whole to determine its hazards, the results of such testing shall be used to determine whether the mixture is hazardous;

(ii) If a mixture has not been tested as a whole to determine whether the mixture is a health hazard, the mixture shall be assumed to present the same health hazards as do the components which comprise one percent (by weight or volume) or greater of the mixture, except that the mixture shall be assumed to present a carcinogenic hazard if it contains a component in concentrations of 0.1 percent or greater which is considered to be a carcinogen under paragraph (d)(4) of this section;

(iii) If a mixture has not been tested as a whole to determine whether the mixture is a physical hazard, the chemical manufacturer, importer, or employer may use whatever scientifically valid data is available to evaluate the physical hazard potential of the mixture; and

(iv) If the employer has evidence to indicate that a component present in the mixture in concentrations of less than one percent (or in the case of carcinogens, less than 0.1 percent) could be released in concentrations which would exceed an established OSHA permissible exposure limit or ACGIH Threshold Limit Value, or could present a health hazard to employees in those concentrations, the mixture shall be assumed to present the same hazard.

(6) Chemical manufacturers, importers, or employers evaluating chemicals shall describe in writing the procedures they use to determine the hazards of the chemical they evaluate. The written procedures are to be made available, upon request, to employees, their designated representatives, the Assistant Secretary and the Director. The written description may be incorporated into the written hazard communication program required under paragraph (e) of this section.

(e) Written Hazard Communication Program

(1) Employers shall develop and implement a written hazard communication program for their workplaces which at least describes how the criteria specified in paragraphs (f), (g), and (h) of this section for labels and other forms of warning, material safety data sheets, and employee information and training will be met, and which also includes the following:

(i) A list of the hazardous chemicals known to be present using an identity that is referenced on the appropriate material safety data sheet (the list may be compiled for the workplace as a whole or for individual work areas);

(ii) The methods the employer will use to inform employees of the hazards of non-routine tasks (for example, the cleaning of reactor vessels), and the hazards associated with chemicals contained in unlabeled pipes in their work areas; and,

(iii) The methods the employer will use to inform any contractor employers with employees working in the employer's workplace of the hazardous chemicals their employees may be exposed to while performing their work, and any suggestions for appropriate protective measures.

(2) The employer may rely on an existing hazard communication program to comply with these requirements, provided that it meets the criteria established in this paragraph (e).

(3) The employer shall make the written hazard communication program available, upon request, to employees, their designated representatives, the Assistant Secretary and the Director, in accordance with the requirements of 29 CFR 1910.20(e).

(f) Labels and Other Forms of Warning

(1) The chemical manufacturer, importer, or distributor shall ensure that each container of hazardous chemicals leaving the workplace is labeled, tagged or marked with the following information:

(i) Identity of the hazardous chemical(s);

(ii) Appropriate hazard warnings; and

(iii) Name and address of the chemical manufacturer, importer, or other responsible party.

(2) Chemical manufacturers, importers, or distributors shall ensure that each container of hazardous chemicals leaving the workplace is labeled, tagged, or marked in accordance with this section in a manner which does not conflict with the requirements of the Hazardous Materials Transportation Act (18 U.S.C. 1801 et seq.) and regulations issued under that Act by the Department of Transportation.

(3) If the hazardous chemical is regulated by OSHA in a substance-specific health standard, the chemical manufacturer, importer, distributor or employer shall ensure that the labels or other forms of warning used are in accordance with the requirements of that standard.

(4) Except as provided in paragraphs (f)(5) and (f)(6) the employer shall ensure that each container of hazardous chemicals in the workplace is labeled, tagged, or marked with the following information:

(i) Identity of the hazardous chemical(s) contained therein; and

(ii) Appropriate hazard warnings.

(5) The employer may use signs, placards, process sheets, batch tickets, operating procedures, or other such written materials in lieu of affixing labels to individual stationary process containers, as long as the alternative method identifies the containers to which it is applicable and conveys the information required by paragraph (f)(4) of this section to be on a label. The written materials shall be readily accessible to the employees in their work area throughout each work shift.

(6) The employer is not required to label portable containers into which hazardous chemicals are transferred from labeled containers, and which are intended only for the immediate use of the employee who performs the transfer.

(7) The employer shall not remove or deface existing labels on incoming containers of hazardous chemicals, unless the container is immediately marked with the required information.

(8) The employer shall ensure that labels or other forms of warning are legible, in English, and prominently displayed on the container, or readily available in the work area throughout each work shift. Employers having employees who speak other languages may add the information in their language to the material presented, as long as the information is presented in English as well.

(9) The chemical manufacturer, importer, distributor or employer need not affix new labels to comply with this section if existing labels already convey the required information.

(g) Material Safety Data Sheets

(1) Chemical manufacturers and importers shall obtain or develop a material safety data sheet for each hazardous chemical they produce or import. Employers shall have a material safety data sheet for each hazardous chemical which they use.

(2) Each material safety data sheet shall be in English and shall contain at least the following information:

(i) The identity used on the label, and, except as provided for in paragraph (f) of this section on trade secrets:

(A) If the hazardous chemical is a single substance, its chemical and common name(s);

(B) If the hazardous chemical is a mixture which has been tested as a whole to determine its hazards, the chemical and common name(s) of the ingredients which contribute to these known hazards, and the common name(s) of the mixture itself; or,

(C) If the hazardous chemical is a mixture which has not been tested as a whole:

(*1*) The chemical and common name(s) of all ingredients which have been determined to be health hazards, and which comprise 1% or greater of the composition, except that chemicals identified as carcinogens under paragraph (d)(4) of this section shall be listed if the concentrations are 0.1% or greater; and,

(*2*) The chemical and common name(s) of all ingredients which have been determined to present a physical hazard when present in the mixture;

(ii) Physical and chemical characteristics of the hazardous chemical (such as vapor pressure, flash point);

(iii) The physical hazards of the hazardous chemical, including the potential for fire, explosion, and reactivity;

(iv) The health hazards of the hazardous chemical including signs and symptoms of exposure, and any medical conditions which are generally recognized as being aggravated by exposure to the chemical;

(v) The primary route(s) of entry;

(vi) The OSHA permissible exposure limit, ACGIH Threshold Limit Value, and any other exposure limit used or recommended by the chemical manufacturer, importer, or employer preparing the material safety data sheet, where available;

(vii) Whether the hazardous chemical is listed in the National Toxicology Program (NTP) *Annual Report on Carcinogens* (latest edition) or has been found to be a potential carcinogen in the International Agency for Research on Cancer (IARC) *Monographs* (latest editions), or by OSHA;

(viii) Any generally applicable precautions for safe handling and use which are known to the chemical manufacturer, importer or employer preparing the material safety data sheet, including appropriate hygienic practices, protective measures during repair and maintenance of contaminated equipment, and procedures for clean-up of spills and leaks;

(ix) Any generally applicable control measures which are known to the chemical manufacturer, importer or employer preparing the material safety data sheet, such as appropriate engineering controls, work practices, or personal protective equipment;

(x) Emergency and first-aid procedures;

(xi) The date of preparation of the material safety data sheet or the last change to it; and,

(xii) The name, address and telephone number of the chemical manufacturer, importer, employer, or other responsible party preparing or distributing the material safety data sheet, who can provide additional information on the hazardous chemical and appropriate emergency procedures, if necessary.

(3) If no relevant information is found for any given category on the material safety data sheet, the chemical manufacturer, importer or employer preparing the material safety data sheet shall mark it to indicate that no applicable information was found.

(4) Where complex mixtures have similar hazards and contents (i.e. the chemical ingredients are essentially the same, but the specific composition varies from mixture to mixture), the chemical manufacturer, importer or employer may prepare one material safety data sheet to apply to all of these similar mixtures.

(5) The chemical manufacturer, importer or employer preparing the material safety data sheet shall ensure that the information recorded accurately reflects the scientific evidence used in making the hazard determination. If the chemical manufacturer, importer or employer becomes newly aware of any significant information regarding the hazards of a chemical, or ways to protect against the hazards, this new information shall be added to the material safety data sheet within three months. If the chemical is not currently being produced or imported the chemical manufacturer or importer shall add the information to the material safety data sheet before the chemical is introduced into the workplace again.

(6) Chemical manufacturers or importers shall ensure that distributors and manufacturing purchasers of hazardous chemicals are provided an appropriate material safety data sheet with their initial shipment, and with the first shipment after a material safety data sheet is updated. The chemical manufacturer or importer shall either provide material safety data sheets with the shipped containers or send them to the manufacturing purchaser prior to or at the time of the shipment. If the material safety data sheet is not provided with the shipment, the manufacturing purchaser shall obtain one from the chemical manufacturer, importer, or distributor as soon as possible.

(7) Distributors shall ensure that material safety data sheets, and updated information, are provided to other distributors and manufacturing purchasers of hazardous chemicals.

(8) The employer shall maintain copies of the required material safety data sheets for each hazardous chemical in the workplace, and shall ensure that they are readily accessible during each work shift to employees when they are in their work area(s).

(9) Material safety data sheets may be kept in any form, including operating procedures, and may be designed to cover groups of hazardous chemicals in a work area where it may be more appropriate to address the hazards of a process rather than individual hazardous chemicals. However, the employer shall ensure that in all cases the required information is provided for each hazardous chemical, and is readily accessible during each work shift to employees when they are in their work area(s).

(10) Material safety data sheets shall also be made readily available, upon request, to designated representatives and to the Assistant Secretary, in accordance with the requirements of 29 CFR 1910.20(e). The Director shall also be given access to material safety data sheets in the same manner.

(h) Employee information and training

Employers shall provide employees with information and training on hazardous chemicals in their work area at the time of their initial assignment, and whenever a new hazard is introduced into their work area.

(1) *Information.* Employees shall be informed of:

(i) The requirements of this section;

(ii) Any operations in their work area where hazardous chemicals are present; and,

(iii) The location and availability of the written hazard communication program, including the required list(s) of hazardous chemicals, and material safety data sheets required by this section.

(2) *Training.* Employee training shall include at least:

(i) Methods and observations that may be used to detect the presence or release of a hazardous chemical in the work area (such as monitoring conducted by the employer, continuous monitoring devices, visual appearance or odor of hazardous chemicals when being released, etc.);

(ii) The physical and health hazards of the chemicals in the work area;

(iii) The measures employees can take to protect themselves from these hazards, including specific procedures the employer has implemented to protect employees from exposure to hazardous chemicals, such as appropriate work practices, emergency procedures, and personal protective equipment to be used; and,

(iv) The details of the hazard communication program developed by the employer, including an explanation of the labeling system and the material safety data sheet, and how employees can obtain and use the appropriate hazard information.

(i) Trade secrets

(1) The chemical manufacturer, importer or employer may withhold the specific chemical identity, including the chemical name and other specific identification of a hazardous chemical, from the material safety data sheet, provided that:

(i) The claim that the information withheld is a trade secret can be supported;

(ii) Information contained in the material safety data sheet concerning the properties and effects of the hazardous chemical is disclosed;

(iii) The material safety data sheet indicates that the specific chemical identity is being withheld as a trade secret; and,

(iv) The specific chemical identity is made available to health professionals, in accordance with the applicable provisions of this paragraph.

(2) Where a treating physician or nurse determines that a medical emergency exists and the specific chemical identity of a hazardous chemical is necessary for emergency or first-aid treatment, the chemical manufacturer, importer, or employer shall immediately disclose the specific chemical identity of a trade secret chemical to that treating physician or nurse, regardless of the existence of a written statement of need or a

confidentiality agreement. The chemical manufacturer, importer, or employer may require a written statement of need and confidentiality agreement, in accordance with the provisions of paragraphs (i) (3) and (4) of this section, as soon as circumstances permit.

(3) In non-emergency situations, a chemical manufacturer, importer, or employer shall, upon request, disclose a specific chemical identity, otherwise permitted to be withheld under paragraph (i)(1) of this section, to a health professional (i.e. physician, industrial hygienist, toxicologist, or epidemiologist) providing medical or other occupational health services to exposed employee(s) if:

(i) the request is in writing;

(ii) The request describes with reasonable detail one or more of the following occupational health needs for the information:

(A) To assess the hazards of the chemicals to which employees will be exposed;

(B) To conduct or assess sampling of the workplace atmosphere to determine employee exposure levels;

(C) To conduct pre-assignment or periodic medical surveillance of exposed employees;

(D) To provide medical treatment to exposed employees;

(E) To select or assess appropriate personal protective equipment for exposed employees;

(F) To design or assess engineering controls or other protective measures for exposed employees; and,

(G) To conduct studies to determine the health effects of exposure.

(iii) The request explains in detail why the disclosure of the specific chemical identity is essential and that, in lieu thereof, the disclosure of the following information would not enable the health professional to provide the occupational health services described in paragraph (ii) of this section:

(A) The properties and effects of the chemical;

(B) Measures for controlling workers' exposure to the chemical;

(C) Methods of monitoring and analyzing worker exposure to the chemical; and,

(D) Methods of diagnosing and treating harmful exposures to the chemical;

(iv) The request includes a description of the procedures to be used to maintain the confidentiality of the disclosed information; and,

(v) The health professional, and the employer or contractor of the health professional's services (i.e., downstream employer, labor organization, or individual employer), agree in a written confidentiality agreement that the health professional will not use the trade secret information for any purpose other than the health need(s) asserted and agree not to release the information under any circumstances other than to OSHA, as provided in paragraph (i)(6) of this section, except as authorized by the terms of the agreement or by the chemical manufacturer, importer, or employer.

(4) The confidentiality agreement authorized by paragraph (i)(3)(iv) of this section:

(i) May restrict the use of the information to the health purposes indicated in the written statement of need;

(ii) May provide for appropriate legal remedies in the event of a breach of the agreement, including stipulation of a reasonable pre-estimate of likely damages; and,

(iii) May not include requirements for the posting of a penalty bond.

(5) Nothing in this standard is meant to preclude the parties from pursuing non-contractual remedies to the extent permitted by law.

(6) If the health professional receiving the trade secret information decides that there is a need to disclose it to OSHA, the chemical manufacturer, importer, or employer who provided the information shall be informed by the health professional prior to, or at the same time as, such disclosure.

(7) If the chemical manufacturer, importer, or employer denies a written request for disclosure of a specific chemical identity, the denial must:

(i) Be provided to the health professional within thirty days of the request;

(ii) Be in writing;

(iii) Include evidence to support the claim that the specific chemical identity is a trade secret;

(iv) State the specific reasons why the request is being denied; and,

(v) Explain in detail how alternative information may satisfy the specific medical or occupational health need without revealing the specific chemical identity.

(8) The health professional whose request for information is denied under paragraph (i)(3) of this section may refer the request and the written denial of the request to OSHA for consideration.

(9) When a health professional refers the denial to OSHA under paragraph (i)(8) of this section, OSHA shall consider the evidence to determine if:

(i) The chemical manufacturer, importer, or employer has supported the claim that the specific chemical identity is a trade secret;

(ii) The health professional has supported the claim that there is a medical or occupational health need for the information; and,

(iii) The health professional has demonstrated adequate means to protect the confidentiality.

(10) (i) If OSHA determines that the specific chemical identity requested under paragraph (i)(3) of this section is not a *bona fide* trade secret, or that it is a trade secret but the requesting health professional has a legitimate medical or occupational health need for the information, has executed a written confidentiality agreement, and has shown adequate means to protect the confidentiality of the information, the chemical manufacturer, importer, or employer will be subject to citation by OSHA.

(ii) If a chemical manufacturer, importer, or employer demonstrates to OSHA that the execution of a confidentiality agreement would not provide sufficient protection against the potential harm from the unauthorized disclosure of a trade secret specific chemical identity, the Assistant Secretary may issue such orders or impose such additional limitations or conditions upon the disclosure of the requested chemical information as may be appropriate to assure that the occupational health services are provided without an undue risk of harm to the chemical manufacturer, importer, or employer.

(11) If, following the issuance of a citation and any protective orders, the chemical manufacturer, importer, or employer continues to withhold the information, the matter is referable to the Occupational Safety and Health Review Commission for enforcement of the citation. In accordance with Commission rules, the Administrative Law Judge may review the citation and supporting documentation *in camera* or issue appropriate protective orders.

(12) Notwithstanding the existence of a trade secret claim, a chemical manufacturer, importer, or employer shall, upon request, disclose to the Assistant Secretary any information which this section requires the chemical manufacturer, importer, or employer to make available. Where there is a trade secret claim, such claim shall be made no later than at the time the information is provided to the Assistant Secretary so that suitable determinations of trade secret status can be made and the necessary protections can be implemented.

(13) Nothing in this paragraph shall be construed as requiring the disclosure under any circumstances of process or percentage of mixture information which is trade secret.

(j) Effective dates

Employers shall be in compliance with this section within the following time periods:

(1) Chemical manufacturers and importers shall label containers of hazardous chemicals leaving their workplaces, and provide material safety data sheets with initial shipments by November 25, 1985.

(2) Distributors shall be in compliance with all provisions of this section applicable to them by November 25, 1985.

(3) Employers shall be in compliance with all provisions of this section by May 25, 1986, including initial training for all current employees.

Appendix A—Health Hazard Definitions (Mandatory)

Although safety hazards related to the physical characteristics of a chemical can be objectively defined in terms of testing requirements (e.g. flammability), health hazard definitions are less precise and more subjective. Health hazards may cause measurable changes in the body—such as decreased pulmonary function. These changes are generally indicated by the occurrence of signs and symptoms in the exposed employees—such as shortness of breath, a non-measurable, subjective feeling. Employees exposed to such hazards must be apprised of both the change in body function and the signs and symptoms that may occur to signal that change.

The determination of occupational health hazards is complicated by the fact that many of the effects or signs and symptoms occur commonly in non-occupationally exposed populations, so that effects of exposure are difficult to separate from normally occurring illnesses. Occasionally, a substance causes an effect that is rarely seen in the population at large, such as angiosarcomas caused by vinyl chloride exposure, thus making it easier to ascertain that the occupational exposure was the primary causative factor. More often, however, the effects are common, such as lung cancer. The situation is further complicated by the fact that most chemicals have not been adequately tested to determine their health hazard potential, and data do not exist to substantiate these effects.

There have been many attempts to categorize effects and to define them in various ways. Generally, the terms "acute" and "chronic" are used to delineate between effects on the basis of severity or duration. "Acute" effects usually occur rapidly as a result of short-term exposures, and are of short duration. "Chronic" effects generally occur as a result of long-term exposure, and are of long duration.

The acute effects referred to most frequently are those defined by the American National Standards Institute (ANSI) standard for Precautionary Labeling of Hazardous Industrial Chemicals (Z129.1-1982)—irritation, corrosivity, sensitization and lethal dose. Although these are important health effects, they do not adequately cover the considerable range of acute effects which may occur as a result of occupational exposure, such as, for example, narcosis.

Similarly, the term chronic effect is often used to cover only carcinogenicity, teratogenicity, and mutagenicity. These effects are obviously a concern in the workplace, but again, do not adequately cover the area of chronic effects, excluding, for example, blood dyscrasias (such as anemia), chronic bronchitis and liver atrophy.

The goal of defining precisely, in measurable terms, every possible health effect that may occur in the workplace as a result of chemical exposures cannot realistically be accomplished. This does not negate the need for employees to be informed of such effects and protected from them.

Appendix B, which is also mandatory, outlines the principles and procedures of hazard assessment.

For purposes of this section, any chemicals which meet any of the following definitions, as determined by the criteria set forth in Appendix B are health hazards:

1. *Carcinogen:* A chemical is considered to be a carcinogen if:

(a) It has been evaluated by the International Agency for Research on Cancer (IARC), and found to be a carcinogen or potential carcinogen; or

(b) It is listed as a carcinogen or potential carcinogen in the *Annual Report on Carcinogens* published by the National Toxicology Program (NTP) (latest edition); or,

(c) It is regulated by OSHA as a carcinogen.

2. *Corrosive:* A chemical that causes visible destruction of, or irreversible alterations in, living tissue by chemical action at the site of contact. For example, a chemical is considered to be corrosive if, when

tested on the intact skin of albino rabbits by the method described by the U.S. Department of Transportation in Appendix A to 49 CFR Part 173, it destroys or changes irreversibly the structure of the tissue at the site of contact following an exposure period of four hours. This term shall not refer to action on inanimate surfaces.

3. *Highly toxic:* A chemical falling within any of the following categories:

(a) A chemical that has a median lethal dose (LD_{50}) of 50 milligrams or less per kilogram of body weight when administered orally to albino rats weighing between 200 and 300 grams each.

(b) A chemical that has a median lethal dose (LD_{50}) of 200 milligrams or less per kilogram of body weight when administered by continuous contact for 24 hours (or less if death occurs within 24 hours) with the bare skin of albino rabbits weighing between two and three kilograms each.

(c) A chemical that has a median lethal concentration (LC_{50}) in air of 200 parts per million by volume or less of gas or vapor, or 2 milligrams per liter or less of mist, fume, or dust, when administered by continuous inhalation for one hour (or less if death occurs within one hour) to albino rats weighing between 200 and 300 grams each.

4. *Irritant:* A chemical, which is not corrosive, but which causes a reversible inflammatory effect on living tissue by chemical action at the site of contact. A chemical is a skin irritant if, when tested on the intact skin of albino rabbits by the methods of 18 CFR 1500.41 for four hours exposure or by other appropriate techniques, it results in an empirical score of five or more. A chemical is an eye irritant if so determined under the procedure listed in 18 CFR 1500.42 or other appropriate techniques.

5. *Sensitizer:* A chemical that causes a substantial proportion of exposed people or animals to develop an allergic reaction in normal tissue after repeated exposure to the chemical.

6. *Toxic:* A chemical falling within any of the following categories:

(a) A chemical that has a median lethal dose (LD_{50}) of more than 50 milligrams per kilogram but not more than 500 milligrams per kilogram of body weight when administered orally to albino rats weighing between 200 and 300 grams each.

(b) A chemical that has a median lethal dose (LD_{50}) of more than 200 milligrams per kilogram but not more than 1,000 milligrams per kilogram of body weight when administered by continuous contact for 24 hours (or less if death occurs within 24 hours) with the bare skin of albino rabbits weighing between two and three kilograms each.

(c) A chemical that has a median lethal concentration (LC_{50}) in air of more than 200 parts per million but not more than 2,000 parts per million by volume of gas or vapor, or more than two milligrams per liter but not more than 20 milligrams per liter of mist, fume, or dust, when administered by continuous inhalation for one hour (or less if death occurs within one hour) to albino rats weighing between 200 and 300 grams each.

7. *Target organ effects:* The following is a target organ categorization of effects which may occur, including examples of signs and symptoms

and chemicals which have been found to cause such effects. These examples are presented to illustrate the range and diversity of effects and hazards found in the workplace, and the broad scope employers must consider in this area, but are not intended to be all-inclusive.

a. Hepatotoxins	Chemicals which produce liver damage
Signs and Symptoms	Jaundice, liver enlargement
Chemicals	Carbon tetrachloride, nitrosamines
b. Nephrotoxins	Chemicals which produce kidney damage
Signs and Symptoms	Edema, proteinuria
Chemicals	Halogenated hydrocarbons, uranium
c. Neurotoxins	Chemicals which produce their primary toxic effects on the nervous system
Signs and Symptoms	Narcosis, behavioral changes, decrease in motor functions
Chemicals	Mercury, carbon disulfide
d. Agents which act on the blood or hematopoietic system	Decrease hemoglobin function, deprive the body tissues of oxygen
Signs and Symptoms	Cyanosis, loss of consciousness
Chemicals	Carbon monoxide, cyanides
e. Agents which damage the lung	Chemicals which irritate or damage the pulmonary tissue
Signs and Symptoms	Cough, tightness in chest, shortness of breath
Chemicals	Silica, asbestos
f. Reproductive toxins	Chemicals which affect the reproductive capabilities, including chromosomal damage (mutations) and effects on fetuses (teratogenesis)
Signs and Symptoms	Birth defects, sterility
Chemicals	Lead, DBCP
g. Cutaneous hazards	Chemicals which affect the dermal layer of the body
Signs and Symptoms	Defatting of the skin, rashes, irritation
Chemicals	Ketones, chlorinated compounds
h. Eye hazards	Chemicals which affect the eye or visual capacity
Signs and Symptoms	Conjunctivitis, corneal damage
Chemicals	Organic solvents, acids

Appendix B—Hazard Determination (Mandatory)

The quality of a hazard communication program is largely dependent upon the adequacy and accuracy of the hazard determination. The hazard determination requirement of this standard is performance-oriented. Chemical manufacturers, importers, and employers evaluating chemicals are not required to follow any specific methods for determining hazards, but they must be able to demonstrate that they have adequately ascertained the hazards of the chemicals produced or imported in accordance with the criteria set forth in this Appendix.

Hazard evaluation is a process which relies heavily on the professional judgment of the evaluator, particularly in the area of chronic hazards. The performance-orientation of the hazard determination does not diminish the duty of the chemical manufacturer, importer or employer to conduct a thorough evaluation, examining all relevant data and producing a scientifically defensible evaluation. For purposes of this standard, the following criteria shall be used in making hazard determinations that meet the requirements of this standard.

1. *Carcinogenicity:* As described in paragraph (d)(4) and Appendix A of this section, a determination by the National Toxicology Program, the International Agency for Research on Cancer, or OSHA that a chemical is a carcinogen or potential carcinogen will be considered conclusive evidence for purposes of this section.

2. *Human data:* Where available, epidemiological studies and case reports of adverse health effects shall be considered in the evaluation.

3. *Animal data:* Human evidence of health effects in exposed populations is generally not available for the majority of chemicals produced or used in the workplace. Therefore, the available results of toxicological testing in animal populations shall be used to predict the health effects that may be experienced by exposed workers. In particular, the definitions of certain acute hazards refer to specific animal testing results (see Appendix A).

4. *Adequacy and reporting of data:* The results of any studies which are designed and conducted according to established scientific principles, and which report statistically significant conclusions regarding the health effects of a chemical, shall be a sufficient basis for a hazard determination and reported on any material safety data sheet. The chemical manufacturer, importer, or employer may also report the results of other scientifically valid studies which tend to refute the findings of hazard.

Appendix C—Information Sources (Advisory)

The following is a list of available data sources which the chemical manufacturer, importer, or employer may wish to consult to evaluate the hazards of chemicals they produce or import:

—Any information in their own company files such as toxicity testing results or illness experience of company employees.

—Any information obtained from the supplier of the chemical, such as material safety data sheets or product safety bulletins.

—Any pertinent information obtained from the following source list (latest editions should be used):

Condensed Chemical Dictionary
Van Nostrand Reinhold Co., 115 Fifth Avenue, New York, NY 10003

The Merck Index: An Encyclopedia of Chemicals and Drugs
Merck and Company, Inc., 128 E. Lincoln Avenue, Rahway, NJ 07065

IARC Monographs on the Evaluation of the Carcinogenic Risk of Chemicals to Man

Geneva: World Health Organization, International Agency for Research on Cancer, 1972–1977. (Multivolume work), 49 Sheridan Street, Albany, New York

Industrial Hygiene and Toxicology, by F. A. Patty
John Wiley & Sons, Inc., New York, NY (Five volumes)

Clinical Toxicology of Commercial Products
Cleason, Gosselin and Hodge

Casarett and Doull's Toxicology; The Basic Science of Poisons
Doull, Klaassen, and Amdur, Macmillan Publishing Co., Inc., New York, NY

Industrial Toxicology, by Alice Hamilton and Harriet L. Hardy
Publishing Sciences Group, Inc., Acton, MA

Toxicology of the Eye, by W. Morton Grant
Charles C. Thomas, 301-327 East Lawrence Avenue, Springfield, IL

Recognition of Health Hazards in Industry
William A. Burgess, John Wiley and Sons, 605 Third Avenue, New York, NY 10158

Chemical Hazards of the Workplace
Nick H. Proctor and James P. Hughes, J. P. Lipincott Company, 6 Winchester Terrace, New York, NY 10022

Handbook of Chemistry and Physics
Chemical Rubber Company, 18901 Cranwood Parkway, Cleveland, OH 44128

Threshold Limit Values for Chemical Substances and Physical Agents in the Workroom Environment with Intended Changes
American Conference of Governmental Industrial Hygienists, 8500 Glenway Avenue, Bldg. D-5, Cincinnati, OH 45211

Note.—The following documents are on sale by the Superintendent of Documents, U.S. Government Printing Office, Washington, D.C. 20402.

Occupational Health Guidelines
NIOSH/OSHA (NIOSH Pub. No. 81–123)

NIOSH/OSHA Pocket Guide to Chemical Hazards
NIOSH Pub. No. 78–210

Registry of Toxic Effects of Chemical Substances
U.S. Department of Health and Human Services, Public Health Service, Center for Disease Control, National Institute for Occupational Safety and Health (NIOSH Pub. No. 80–102)

The Industrial Environment—Its Evaluation and Control
U.S. Department of Health and Human Services, Public Health Service, Center for Disease Control, National Institute for Occupational Safety and Health (NIOSH Pub. No. 74–117)

Miscellaneous Documents—National Institute for Occupational Safety and Health

1. Criteria for a recommended standard
   * * * Occupational Exposure to "——"
2. Special Hazard Reviews

3. Occupational Hazard Assessment
4. Current Intelligence Bulletins

Bibliographic Data Bases

Service Provider and File Name

Bibliographic Retrieval Services (BRS), Corporation Park, Bldg. 702, Scotia, New York 12302

AGRICOLA
BIOSIS PREVIEWS
CA CONDENSATES
CA SEARCH
DRUG INFORMATION
MEDLARS
MEDOC
NTIS
POLLUTION ABSTRACTS
SCIENCE CITATION INDEX
SSIE

Lockheed—DIALOG, Lockheed Missiles & Space Company, Inc., P.O. Box 44481, San Francisco, CA 94144

AGRICOLA
BIOSIS PREV. 1972–PRESENT
BIOSIS PREV. 1969–71
CA CONDENSATES 1970–71
CA SEARCH 1972–76
CA SEARCH 1977–PRESENT
CHEMNAME
CONFERENCE PAPERS INDEX
FOOD SCIENCE & TECH. ABSTR.
FOOD ADLIBRA
INTL. PHARMACEUTICAL ABSTR.
NTIS
POLLUTION ABSTRACTS
SCISEARCH 1978–PRESENT
SCISEARCH 1974–77
SSIE CURRENT RESEARCH

SDC—ORBIT, SDC Search Service, Department No. 2230, Pasadena, CA 91051

AGRICOLA
BIOCODES
BIOSIS/BIO6973
CAS6771/CAS7276
CAS77
CHEMDEX
CONFERENCE
ENVIROLINE
LABORDOC
NTIS
POLLUTION
SSIE

Chemical Information System (CIS), Chemical Information Systems Inc., 7215 Yorke Road, Baltimore, MD 21212
- Structure & Nomenclature Search System
- Acute Toxicity (RTECS)
- Clinical Toxicology of Commercial Products
- Oil and Hazardous Materials Technical Assistance Data System

National Library of Medicine, Department of Health and Human Services, Public Health Service, National Institutes of Health, Bethesda, MD 20209
- Toxicology Data Bank (TDB)
- MEDLIN
- TOXLINE
- CANCERLIT
- RTECS

Appendix 2

OSHA 29 CFR 1910 HEALTH AND SAFETY STANDARDS: OCCUPATIONAL EXPOSURE TO TOXIC SUBSTANCES IN LABORATORIES; PROPOSED RULE

(a) Scope and application

(1) This section shall apply to all employers engaged in the laboratory use of toxic substances as defined below.

(2) Where this section applies, it shall supersede, for laboratories, the requirements of all other OSHA health standards in 29 CFR Part 1910, Subpart Z, except the requirement to limit employee exposure to the specified permissible exposure limits.

(3) For any OSHA health standard which is promulgated after the effective date of this section, only the requirement to limit employee exposure to the specific permissible exposure limit shall apply for laboratories, unless that particular standard states otherwise.

(4) This section shall not apply to uses of toxic substances which do not meet the definition of laboratory use, and in such cases, the employer shall comply with the relevant standard in 29 CFR Part 1910, Subpart Z, even if such use occurs in a laboratory.

(b) Definitions

"Assistant Secretary" means the Assistant Secretary of Labor for Occupational Safety and Health, U.S. Department of Labor, or designee.

"Carcinogen" or "potential carcinogen" means any substance which meets one of the following criteria: (1) Is regulated by OSHA as a carcinogen; or (2) is identified by the International Agency for Research on Cancer (IARC) or the National Toxicology Program (NTP) as a carcinogen or potential carcinogen.

"Chemical Hygiene Officer" means an employee who is designated by the employer, and who is qualified by training and experience, to provide technical guidance in the development and implementation of the provisions of the Chemical Hygiene Plan. This definition is not intended to place limitations on the position description or job classification that the designated individual shall hold within the employer's organizational structure.

"Chemical Hygiene Plan" means a reasonable written program developed and implemented by the employer which sets forth procedures, equipment, personal protective equipment and work practices that are capable of protecting employees from the health hazards presented by toxic substances used in that particular workplace and which meets the requirements of Paragraph (d) of this section.

"Closed system" means a device such as a glove box or other system which physically encloses an operation or procedure involving the laboratory use of toxic substances; is constructed and maintained to provide a physical separation between the employee and the substances used in the workplace; is designed to prevent vapors from escaping from the closed system into the laboratory environment; and allows manipulation of chemicals to be conducted in the enclosure by the use of remote controls or gloves which are physically attached and sealed to the enclosure.

"Emergency" means any occurrence such as, but not limited to, equipment failure, rupture of containers or failure of control equipment which results in an uncontrolled release of hazardous amounts of a toxic substance into the workplace.

"Exposure evaluation" means an assessment of the conditions present during a laboratory operation, procedure or activity for the purpose of determining if an employee has or may have been overexposed to a toxic substance.

"Laboratory" means a facility where the "laboratory use of toxic substances" occurs. It is a workplace where relatively small quantities of toxic substances are used on a nonproduction basis.

"Laboratory scale" means work with substances in which the containers used for reactions, transfers, and other handling of substances are designed for manual use, being small enough to be easily and safely manipulated by one person. "Laboratory scale" excludes those workplaces whose function is to produce commercial quantities of materials.

"Laboratory-type hood" means a device located in a laboratory, enclosed on five sides with a moveable sash or fixed partial enclosure on the remaining side; constructed and maintained to draw air from the laboratory and to prevent or minimize the escape of air contaminants into the laboratory; and allows chemical manipulations to be conducted in the enclosure without insertion of any portion of the employee's body other than hand and arms.

Walk-in hoods with adjustable sashes meet the above definition provided that the sashes are adjusted during use so that the airflow and the exhaust of air contaminants are not compromised and employees do not work inside the enclosure during the release of airborne toxic substances.

"Laboratory use of toxic substances" means handling or use of such substances in which all of the following conditions are met: (1) chemical manipulations are carried out on a "Laboratory scale"; (2) multiple chemical procedures and/or chemicals are in use in a single room; and (3) protective laboratory practices which may include the use of appropriate equipment are available and in common use to minimize the potential for employee overexposure to toxic substances.

"Medical consultation" means a consultation which takes place between an employee and a licensed physician for the purpose of determining what medical examinations or procedures, if any, are appropriate in cases where an "exposure evaluation" has determined that an "overexposure" may have taken place.

"Overexposure" means an employee exposure in excess of the permissible exposure limits (PELs) for an OSHA-regulated substance.

"Protective laboratory practices and equipment" means those laboratory procedures, practices and equipment accepted by laboratory health and safety experts as effective, or that the employer can show to be effective, in minimizing the potential for employee exposure to toxic substances.

"Regulated area" means a laboratory, an area of a laboratory or device such as a laboratory hood for which access is limited to persons who are aware of the hazards of the substances in use and the precautions that are necessary.

"Toxic substance" means any substance which is: (1) Regulated by OSHA in 29 CFR Part 1910, Subpart Z or (2) is found to be a carcinogen or potential carcinogen as defined in this paragraph.

(c) Permissible exposure limits

For laboratory uses of OSHA-regulated substances, the employer shall assure that laboratory employees' exposures to such substances do not exceed the permissible exposure limits specified in 29 CFR Part 1910, Subpart Z. In addition, prohibition of eye and skin contact where specified by any standard in this Subpart shall be observed.

(d) Chemical hygiene plan

(1) The employer shall develop and carry out the provisions of a reasonable written Chemical Hygiene Plan which is capable of protecting employees from health hazards associated with toxic substances in that laboratory and is capable of keeping exposures below the limits specified in paragraph (c) of this section. (Appendix A provides guidance to assist employers in the development of the Chemical Hygiene Plan.) The Chemical Hygiene Plan shall be readily available to employees and upon request to the Assistant Secretary.

(2) The Chemical Hygiene Plan shall include each of the following elements and shall indicate specific measures to ensure laboratory employee protection:

(i) Standard operating procedures to be followed when laboratory work involves the use of toxic substances;

(ii) Criteria that the employer will use to determine and implement control measures to reduce employee exposure to toxic substances including engineering controls, the use of personal protective equipment and hygiene practices;

(iii) A requirement that fume hoods and other protective equipment are functioning properly and specific measures that shall be taken to ensure proper and adequate performance of such equipment; and

(iv) Information and training procedures to ensure that employees are apprised of the potential health hazards in their workplace and the measures to be taken with regard to employee protection. The employer shall ensure that each employee is informed of the following:

(A) The existence, location and availability of the employer's Chemical Hygiene plan;

(B) This standard and its appendices;

(C) The selection, proper use and limitations of protective equipment and clothing to effectively minimize employee exposure during the laboratory use of toxic substances, where such equipment and clothing are appropriate;

(D) The permissible exposure limits for OSHA-regulated substances, or the threshold limit values recommended by the American Conference of Governmental Industrial Hygienists where there is no applicable OSHA standard;

(E) Procedures which shall be followed in the event of an emergency, including the location and proper use of available emergency equipment; and

(F) The availability of reference material on the hazards and safe handling of toxic substances including, but not limited to, any information such as material safety data sheets that may be available from the chemical supplier.

(v) The circumstances under which a particular laboratory operation, procedure or activity shall require prior approval from the employer or the employer's designee before implementation;

(vi) Provisions for an exposure evaluation for employees who, as a consequence of a laboratory operation, procedure or activity, reasonably suspect or believe they have sustained an overexposure to a toxic substance. The exposure evaluation shall be conducted by the Chemical Hygiene Officer or other person qualified by training and experience.

The employer shall also consider whether it is appropriate to provide an exposure evaluation in the case of a possible exposure in excess of an ACGIH TLV for a substance which has no associated OSHA PEL.

(vii) Provisions for a medical consultation for any employee whenever an exposure evaluation indicates that the employee is likely to have sustained an overexposure to a toxic substance. The employer shall ensure that the medical consultation is provided without cost to the employee and includes a review by the consulting physician of the employee's exposure evaluation and an opportunity for the physician to confer with the employee as necessary to determine if medical examinations and/or medical surveillance are appropriate.

(viii) An opportunity for the employee to receive without cost any medical examinations and/or medical surveillance recommended by the consulting physician as a result of the medical consultation in accordance with the prescribed time interval. The physician shall furnish the employer and employee with a written opinion. The written opinion obtained by the employer shall not reveal specific findings and diagnoses unrelated to occupational exposure.

(ix) Designation of personnel responsible for implementation of the Chemical Hygiene Plan including the assignment of a Chemical Hygiene Officer and, if appropriate, establishment of Chemical Hygiene Committee; and

(x) Provisions for additional employee protection for work with carcinogens or potential carcinogens as defined herein. Such provisions shall include, as a minimum, the following elements:

(A) Establishment of a regulated area;

(B) Requirement that such work be conducted in a properly functioning laboratory type hood, closed system or other device which provides equivalent employee protection;

(C) Specification of procedures for the safe removal of contaminated waste;

(D) Specification of personal hygiene practices to be exercised within and immediately upon exiting a regulated area;

(E) Specification of appropriate procedures to be employed to protect vacuum lines and vacuum pumps from contamination; and

(F) Appropriate protective apparel to be worn while working within the regulated area.

(3) The employer shall review and evaluate the effectiveness of the Chemical Hygiene Plan at least annually and update it as necessary.

(e) Use of respirators

Where the use of respirators is necessary to maintain exposure below permissible exposure limits, the employer shall provide, at no cost to the employee, the proper respiratory equipment. Respirators shall be selected and used in accordance with the requirements of 29 CFR 1910.134.

(f) Recordkeeping

(1) The employer shall establish and maintain for each employee an accurate record of any medical consultation and examinations including tests and written opinions conducted in accordance with paragraphs (d)(2)(vii) and (d)(2)(viii) of this section. This record shall also contain the results of the exposure evaluation including an estimate of the possible extent of overexposure.

(2) The employer shall assure that such records are kept, transferred, and made available in accordance with 29 CFR 1910.20.

(g) Dates

(1) *Effective date.* This section shall become effective 90 days after publication.

(2) *Startup dates.*

(i) Employers shall have completed development of a reasonable written Chemical Hygiene Plan and commenced carrying out its provisions as specified by paragraph (d) of this section no later than one year from the publication date of the standard.

(ii) The provisions of paragraph (a)(2) of this section shall not take effect until the employer has completed the development of a reasonable written Chemical Hygiene Plan as specified by paragraph (d) of this section.

(h) Appendices

The information contained in the appendices is not intended, by itself, to create any additional obligations not otherwise imposed or to detract from any existing obligation.

Appendix A to National Research Council Recommendations Concerning Chemical Hygiene in Laboratories (Non-Mandatory)

Foreword

As guidance for each employer's development of an appropriate laboratory Chemical Hygiene Plan, the following non-mandatory recommendations are provided. They were extracted from "Prudent Practices for Handling Hazardous Chemicals in Laboratories" (referred to below as "Prudent Practices"), which was published in 1981 by the National Research Council and is available from the National Academy Press, 2101 Constitution Ave., NW., Washington, DC 20418.

"Prudent Practices" is cited because of its wide distribution and acceptance and because of its preparation by members of the laboratory community through the sponsorship of the National Research Council. However, none of the recommendations given here will modify any of requirements of the laboratory standard. This Appendix merely presents pertinent recommendations from "Prudent Practices," organized into a form convenient for quick reference during operation of a laboratory facility and during development and application of a Chemical Hygiene Plan. Users of this appendix should consult "Prudent Practices" for a more extended presentation and justification for each recommendation.

"Prudent Practices" deals with both safety and chemical hazards while the proposed laboratory standard is concerned only with toxic hazards. Therefore, only those recommendations directed primarily toward control of toxic exposures are cited in this appendix, with the term "chemical hygiene" being substituted for the word "safety." However, since conditions producing or threatening physical injury often pose toxic risks as well, page references concerning major categories of safety hazards in the laboratory are given in section F.

The recommendations from "Prudent Practices" have been paraphrased, combined, or otherwise reorganized, and headings have been added. However, their sense has not been changed.

Corresponding Sections of the Proposed Standard and This Appendix

The following table is given for the convenience of those who are developing a Chemical Hygiene Plan which will satisfy the requirements of paragraph (d) of the standard. It indicates those sections of this appendix which are most pertinent to each of the sections of paragraph (d).

Paragraph and topic in proposed laboratory standard	Relevant appendix section
(d)(2)(I) Standard operating procedures for handling toxic chemicals	C, D, E
(d)(2)(ii) Criteria to be used for implementation of measures to reduce exposures	D
(d)(2)(iii) Fume hood performance	C4b
(d)(2)(iv) Provision of training and information (including emergency procedures)	D10, D9
(d)(2)(v) Requirements for prior approval of laboratory activities	E2b, E4b
(d)(2)(vi) Exposure evaluation	D3, C4h
(d)(2)(vii) Medical consultation	D5, E4f
(d)(2)(viii) Provision for medical follow-up	D5
(d)(2)(ix) Chemical hygiene responsibilities	B
(d)(2)(x) Special precautions for work with carcinogens	E3, E4
(A) Establishment of a regulated area	E3c, E4a
(B) Use of hoods or closed systems	E4a
(C) Waste removal	E3g, E4l, E5e
(D) Personal hygiene practices	E3d, E4d
(E) Protection of vacuum lines and pumps	E4c
(F) Use of protective apparel	E3d

In this appendix, those recommendations directed primarily at administrators and supervisors are given in Sections A–D. Those recommendations of primary concern to employees who are actually handling laboratory chemicals are given in section E. (References to page numbers in "Prudent Practices" are given in parentheses.)

A. General Principles for Work With Laboratory Chemicals

In addition to the more detailed recommendations listed below in sections B–E, "Prudent Practices" expresses certain general principles, including the following:

1. *It is Prudent to Minimize all Chemical Exposures.* Because few laboratory chemicals are without hazards, general precautions for handling all laboratory chemicals should be adopted, rather than specific guidelines for particular chemicals (2, 10). Skin contact with chemicals should be avoided as a cardinal rule (198).

2. *Avoid Underestimation of Risk.* Even for substances of no known significant hazard, exposure should be minimized; for work with substances which present special hazards, special precautions should be taken (10, 37, 38). One should assume that any mixture will be more toxic than its most toxic component (30, 103) and that all substances of unknown toxicity are toxic (3, 34).

3. *Provide Adequate Ventilation.* The best way to prevent exposure to airborne substances is to prevent their escape into the working atmosphere by use of hoods and other ventilation devices (32, 198).

4. *Institute a Chemical Hygiene Program.* A mandatory chemical hygiene program designed to minimize exposures is needed; it should be a regular, continuing effort, not merely a standby or short-term activity (6, 11). Its recommendations should be followed in academic teaching laboratories as well as by full-time laboratory workers (13).

5. *Observe the PELs, TLVs.* The Permissible Exposure Limits of OSHA and the Threshold Limit Values of the American Conference of Governmental Industrial Hygienists should not be exceeded (13).

B. Chemical Hygiene Responsibilities

Responsibility for chemical hygiene rests at all levels (6, 11, 21) including the:

1. *Chief Executive Officer,* who has ultimate responsibility for chemical hygiene within the institution and must, with other administrators, provide continuing support for institutional chemical hygiene (7, 11).

2. *Supervisor of the Department or other Administrative Unit,* who is responsible for chemical hygiene in that unit (7).

3. *Chemical Hygiene Officer(s),* whose appointment is essential (7) and who must:

(a) work with administrators and other employees to develop and implement appropriate chemical hygiene policies and practices (7);

(b) monitor procurement, use, and disposal of chemicals used in the lab (8);

(c) see that appropriate audits are maintained (8);

(d) help project directors develop precautions and adequate facilities (10);

(e) know the current legal requirements concerning regulated substances (50); and

(f) seek ways to improve the chemical hygiene program (8, 11).

4. *Laboratory Supervisor,* who has overall responsibility for chemical hygiene in the laboratory (21) including responsibility to:

(a) ensure that workers know and follow the chemical hygiene rules, that protective equipment is available and in working order, and that appropriate training has been provided (21, 22);

(b) provided regular, formal chemical hygiene and housekeeping inspections including routine inspections of emergency equipment (21, 171);

(c) know the current legal requirements concerning regulated substances (50, 231);

(d) determine the required levels of protective apparel and equipment (156, 160, 162); and

(e) ensure that facilities and training for use of any material being ordered are adequate (215).

5. *Project Director or Director of other Specific Operation,* who has primary responsibility for chemical hygiene procedures for that operation (7).

6. *Laboratory Worker,* who is responsible for:

(a) planning and conducting each operation in accordance with the institutional chemical hygiene procedures (7, 21, 22, 230); and

(b) developing good personal chemical hygiene habits (22).

C. The Laboratory Facility

1. *Design.* The laboratory facility should have:

(a) two exits for each laboratory (225):

(b) an appropriate general ventilation system (see C4 below) with air intakes and exhausts located so as to avoid intake of contaminated air (194);

(c) adequate, well-ventilated stockrooms/storerooms (218, 219);

(d) laboratory hoods and sinks (12, 162);

(e) other safety equipment including eyewash fountains and drench showers (162, 169); and

(f) arrangements for waste disposal (12, 240).

2. *Maintenance.* Chemical-hygiene-related equipment (hoods, incinerator, etc.) should undergo continuing appraisal and be modified if inadequate (11, 12).

3. *Usage.* The work conducted (10) and its scale (12) must be appropriate to the physical facilities available and, especially, to the quality of ventilation (13).

4. *Ventilation.*

(a) General laboratory ventilation. This system should: provide a source of air for breathing and for input to local ventilation devices (199); it should not be relied on for protection from toxic substances released into the laboratory (198); ensure that laboratory air is continually replaced, preventing increase of air concentrations of toxic substances during the working day (194); direct air flow into the laboratory from non-laboratory areas and out to the exterior of the building (194).

(b) Hoods. A laboratory hood with 2.5 feet of hood space per person should be provided for every 2 workers if they spend most of their time working with chemicals (199); each hood should have a continuous monitoring device to allow convenient confirmation of adequate hood performance before use (200, 209). If this is not possible, work with substance of unknown toxicity should be avoided (13) or other types of local ventilation devices should be provided (199). See pp. 201–206 for a discussion of hood design, construction and evaluation.

(c) Other local ventilation devices. Ventilated storage cabinets, canopy hoods, snorkels, etc. should be provided as needed (199). Each canopy hood and snorkel should have a separate exhaust duct (207).

(d) Special ventilation areas. Exhaust air from glove boxes and isolation rooms should be passed through scrubbers or other treatment before release into the regular exhaust system (208). Cold rooms and warm rooms should have provisions for rapid escape and for escape in the event of electrical failure (209).

(e) Modifications. Any alteration of the ventilation system should be made only if thorough testing indicates that worker protection from airborne toxic substances will continue to be adequate (12, 193, 204).

(f) Performance. Rate: 4–12 room air changes/hour is normally adequate general ventilation if local exhaust systems such as hoods are used as the primary method of control (194).

(g) Quality. General air flow should not be turbulent and should be relatively uniform throughout the laboratory, with no high velocity or static areas (194, 195); airflow into and within the hood should not be excessively turbulent (200); hood face velocity should be adequate (typically 60–100 lfm) (200, 204).

(h) Evaluation. Quality and quantity of ventilation should be evaluated on installation (202), regularly monitored at least every 3 months (6, 12, 14, 195), and reevaluated whenever a change in local ventilation devices is made (12, 195, 207). See pp. 195–198 for methods of evaluation and for calculation of estimated airborne contaminant concentrations.

D. Components of the Chemical Hygiene Plan

1. *Basic Rules and Procedures.* (Recommendations for these are given in section E, below.)

2. *Chemical Procurement, Distribution, and Storage.*

(a) Procurement. Before a substance is received, information on proper handling, storage, and disposal should be known to those who will be involved (215, 216). No container should be accepted without an adequate identifying label (216). All substances should be preferably received in a central location (216).

(b) Stockrooms/storerooms. Toxic substances should be segregated in a well-identified area with local exhaust ventilation (221). Chemicals which are highly toxic (227) or other chemicals whose containers have been opened should be in unbreakable secondary containers (219). Stored chemicals should be examined periodically (at least annually) for replacement, deterioration, and container integrity (218–19).

Stockrooms/storerooms should not be used as preparation or repackaging areas, should be open during normal working hours, and should be controlled by one person (219).

(c) Distribution. When chemicals are hand carried, the container should be placed in an outside container or bucket. Freight-only elevators should be used if possible (223).

(d) Laboratory storage. Amounts permitted should be as small as practical. Storage on bench tops and in hoods is inadvisable. Exposure to heat or direct sunlight should be avoided. Periodic inventories should be conducted, with unneeded items being discarded or returned to the storeroom/stockroom (225–6, 229).

3. *Environmental Monitoring.* Regular instrumental monitoring of airborne concentrations is not usually justified or practical in laboratories but may be appropriate when testing or redesigning hoods or other ventilation devices (12) or when a highly toxic substance is stored or used regularly (e.g., 3 times/week) (13).

4. *Housekeeping, Maintenance, and Inspection.*

(a) Cleaning. Floors should be cleaned regularly (24).

(b) Inspections. Formal housekeeping and chemical hygiene inspections should be held at least quarterly (6, 21) for units which have frequent personnel changes and semiannually for others; informal inspections should be continual (21).

(c) Maintenance. Eye wash fountains should be inspected at intervals of not less than 3 months (6). Respirators for routine use should be inspected periodically by the laboratory supervisor (169). Safety showers should be tested routinely (169). Other safety equipment should be inspected regularly (e.g., every 3–6 months) (6, 24, 171). Procedures to prevent restarting of out-of-service equipment should be established (25).

(d) Passageways. Stairways and hallways should not be used as storage areas (24). Access to exits, emergency equipment, and utility controls should never be blocked (24).

5. *Medical Program.*

(a) Compliance with regulations. Regular medical surveillance should be established to the extent required by regulations (12).

(b) Routine surveillance. Anyone whose work involves regular and frequent handling of toxicologically significant quantities of a chemical should consult a qualified physician to determine on an individual basis whether a regular schedule of medical surveillance is desirable (11, 50).

(c) First aid. Personnel trained in first aid should be available during working hours and an emergency room with medical personnel should be nearby (173). See pp. 176–178 for description of some emergency first aid procedures.

6. *Protective Apparel and Equipment.* These should include for each laboratory:

(a) protective apparel compatible with the required level of performance for substances being handled (158–161);

(b) an easily accessible drench-type safety shower (162, 169);

(c) an eyewash fountain (162);

(d) a fire extinguisher (162–164);

(e) access to a nearby respirator (164–9), fire alarm and telephone for emergency use (162); and

(f) other items designated by the laboratory supervisor (156, 160).

7. *Records.*

(a) Accident records should be written and retained (174).

(b) Chemical Hygiene Plan records should document that the facilities and precautions were compatible with current knowledge and regulations (7).

(c) Inventory and usage records for high-risk substances should be kept as specified in sections E3e below.

(d) Medical records should be retained by the institution in accordance with the requirements of state and federal regulations (12).

8. *Signs and labels.* Prominent signs and labels of the following types should be posted:

(a) emergency telephone numbers of emergency personnel/facilities, supervisors, and laboratory workers (28);

(b) identity labels, showing contents of containers (including waste receptacles) and associated hazards (27, 48);

(c) location signs for safety showers, eyewash stations, other safety and first aid equipment, exits (27) and areas where food and beverage consumption and storage are permitted (24); and

(d) warnings at areas or equipment where special or unusual hazards exist (27).

9. *Spills and Accidents.*

(a) A written emergency plan should be established and communicated to all personnel; it should include procedures for ventilation failure (200), evacuation, medical care, reporting, and drills (172).

(b) There should be an alarm system to alert people in all parts of the facility including isolation areas such as cold rooms (172).

(c) A spill control policy should be developed and should include consideration of prevention, containment, cleanup, and reporting (175).

(d) All accidents or near accidents should be carefully analyzed with the results distributed to all who might benefit (8, 28).

10. *Training and Information Program.*

(a) Aim: To assure that all individuals at risk are adequately informed about the work in the laboratory, its risks, and what to do if an accident occurs (5, 15).

(b) Emergency and Personal Protection Training: Every laboratory worker should know the location and proper use of available protective apparel and equipment (154, 169).

Some of the full-time personnel of the laboratory should be trained in the proper use of emergency equipment and procedures (6).

Such training as well as first aid instruction should be available to (154) and encouraged for (176) everyone who might need it.

(c) Receiving and stockroom/storeroom personnel should know about hazards, handling equipment, protective apparel, and relevant regulations (217).

(d) Frequency of Training: The training and education program should be a regular, continuing activity—not simply an annual presentation (15).

(e) Literature/Consultation: Literature and consulting advice concerning chemical hygiene should be readily available to laboratory personnel, who should be encouraged to use these information resources (14).

11. *Waste Disposal Program.*

(a) Aim: To assure that minimal harm to people, other organisms, and the environment will result from the disposal of waste laboratory chemicals (5).

(b) Content (14, 232, 233, 240): The waste disposal program should specify how waste is to be collected, segregated, stored, and transported and include consideration of what materials can be incinerated. Transport from the institution must be in accordance with DOT regulations (244).

(c) Discarding Chemical Stocks: Unlabeled containers of chemicals and solutions should undergo prompt disposal; if partially used, they should not be opened (24, 27).

Before a worker's employment in the laboratory ends, chemicals for which that person was responsible should be discarded or returned to storage (226).

(d) Frequency of Disposal: Waste should be removed from laboratories to a central waste storage area at least once per week and from the central waste storage area at regular intervals (14).

(e) Method of Disposal: Incineration in an environmentally acceptable manner is the most practical disposal method for combustible laboratory waste (14, 238, 241).

Indiscriminate disposal by pouring waste chemicals down the drain (14, 231, 242) or adding them to mixed refuse for landfill burial is unacceptable (14).

Hoods should not be used as a means of disposal for volatile chemicals (40, 200).

Disposal by recycling (233, 243) or chemical decontamination (40, 230) should be used when possible.

E. Basic Rules and Procedures for Working With Chemicals

The Chemical Hygiene Plan should require that laboratory workers know and follow its rules and procedures. In addition to the procedures of the sub programs mentioned above, these should include the rules listed below.

1. *General Rules.* The following are to be used for essentially all laboratory work with chemicals:

(a) Accidents and Spills:

Eye Contact: Promptly flush eyes with water for a prolonged period (15 minutes) and seek medical attention (33, 172).

Ingestion: Encourage the victim to drink large amounts of water (178).

Skin Contact: Promptly flush the affected area with water (33, 172, 178) and remove any contaminated clothing (172, 178). If symptoms persist after washing, seek medical attention (33).

Clean-up: Promptly clean up spills, using appropriate protective apparel and equipment and proper disposal (24, 33). See pp. 233–237 for specific clean-up recommendations.

(b) Avoidance of "Routine" Exposure: Develop and encourage safe habits (23); avoid unnecessary exposure to chemicals by any route (23);

Do not smell or taste chemicals (32). Vent apparatus which may discharge toxic chemicals (vacuum pumps, distillation columns, etc.) into local exhaust devices (199).

Inspect gloves (157) and test gloves boxes (208) before use.

Do not allow release of toxic substances in cold rooms and warm rooms, since these have contained recirculated atmospheres (209).

(c) Choice of chemicals: Use only those chemicals for which the quality of the available ventilation system is appropriate (13).

(d) Eating, smoking, etc.: Avoid eating, drinking, smoking, gum chewing, or application of cosmetics in areas where laboratory chemicals are present (22, 24, 32, 40); wash hands before conducting these activities (23, 24).

Avoid storage, handling or consumption of food or beverages in storage areas, refrigerators, glassware or utensils which are also used for laboratory operations (23, 24, 226).

(e) Equipment and Glassware: Handle and store laboratory glassware with care to avoid damage; do not use damaged glassware (25). Use extra care with Dewar flasks and other evacuated glass apparatus; shield or wrap them to contain chemicals and fragments should implosion occur (25). Use equipment only for its designed purpose (23, 26).

(f) Exiting: Wash areas of exposed skin well before leaving the laboratory (23).

(g) Horseplay: Avoid practical jokes or other behavior which might confuse, startle or distract another worker (23).

(h) Mouth Suction: Do not use mouth suction for pipeting or starting a siphon (23, 32).

(i) Personal Apparel: Confine long hair and loose clothing (23, 158). Wear shoes at all times in the laboratory but do not wear sandals, perforated shoes, or sneakers (158).

(j) Personal Housekeeping: Keep the work area clean and uncluttered, with chemicals and equipment being properly labeled and stored; clean up the work area on completion of an operation or at the end of each day (24).

(k) Personal Protection: Assure that appropriate eye protection (154–156) is worn by all persons, including visitors, where chemicals are stored or handled (22, 23, 33, 154).

Wear appropriate gloves when the potential for contact with toxic materials exists (157); inspect the gloves before each use, wash them before removal, and replace them periodically (157). (A table of resistance to chemicals of common glove materials is given on p. 159.)

Use appropriate (164–168) respiratory equipment when air contaminant concentrations are not sufficiently restricted by engineering controls (164–5), inspecting the respirator before use (169).

Use any other protective and emergency apparel and equipment as appropriate (22, 157–162).

Avoid use of contact lenses in the laboratory unless necessary; if they are used, inform supervisor so special precautions can be taken (155).

Remove laboratory coats immediately on significant contamination (161).

(l) Planning: Seek information and advice about hazards (7), plan appropriate protective procedures, and plan positioning of equipment before beginning any new operation (22, 23).

(m) Unattended Operations: Leave lights on, place an appropriate sign on the door, and provide for containment of toxic substances in the event of failure of a utility service (such as cooling water) to an unattended operation (27, 128).

(n) Use of Hood: Use the hood for operations which might result in release of toxic chemical vapors or dust (198–9).

As a rule of thumb, use a hood or other local ventilation device when working with any appreciably volatile substance with a TLV of less than 50 ppm (13).

Confirm adequate hood performance before use; keep hood closed at all times except when adjustments within the hood are being made (200); keep materials stored in hoods to a minimum and do not allow them to block vents or air flow (200).

Leave the hood "on" when it is not in active use if toxic substances are stored in it or if it is uncertain whether adequate general laboratory ventilation will be maintained when it is "off" (200).

(o) Vigilance: Be alert to unsafe conditions and see that they are corrected when detected (22).

(p) Waste Disposal: Assure that the plan for each laboratory operation includes plans and training for waste disposal (230).

Deposit chemical waste in appropriately labeled receptacles and follow all other waste disposal procedures of the Chemical Hygiene Plan (22, 24).

Do not discharge into the sewer concentrated acids or bases (231); highly toxic, malodorous, or lachrymatory substances (231); or any substances which might interfere with the biological activity of waste water treatment plants, create fire or explosion hazards, cause structural damage or obstruct flow (242).

(q) Working Alone: Avoid working alone in a building; do not work alone in a laboratory if the procedures being conducted are hazardous (28).

2. *Working with Allergens and Embryotoxins.*

(a) Allergens (examples: diazomethane, isocyanates, bichromates): Wear suitable gloves to prevent hand contact with allergens or substances of unknown allergenic activity (35).

(b) Embryotoxins (34–5) (examples: organomercurials, lead compounds, formamide): If you are a woman of childbearing age, handle these substances only in a hood whose satisfactory performance has been confirmed, using appropriate protective apparel (especially gloves) to prevent skin contact.

Review each use of these materials with the research supervisor and review continuing uses annually or whenever a procedural change is made.

Store these substances, properly labeled, in an adequately ventilated area in an unbreakable secondary container.

Notify supervisors of all incidents of exposure or spills; consult a qualified physician when appropriate.

3. *Work with Chemicals of Moderate Chronic or High Acute Toxicity* (examples: diisopropylfluorophosphate (41), hydrofluoric acid (43), hydrogen cyanide (45)):

Supplemental rules to be followed in addition to those mentioned above (Procedure B of "Prudent Practices," pp. 39–41):

(a) Aim: To minimize exposure to these toxic substances by any route using all reasonable precautions (39).

(b) Applicability: These precautions are appropriate for substances with moderate chronic or high acute toxicity used in significant quantities (39).

(c) Location: Use and store these substances only in areas of restricted access with special warning signs (40, 229).

Always use a hood (previously evaluated to confirm adequate performance with a face velocity of at least 60 linear feet per minute) (40) or other containment device for procedures which may result in the generation of aerosols or vapors containing the substance (39); trap released vapors to prevent their discharge with the hood exhaust (40).

(d) Personal Protection: Always avoid skin contact by use of gloves and long sleeves (and other protective apparel as appropriate) (39). Always wash hands and arms immediately after working with these materials (40).

(e) Records: Maintain records of the amounts of these materials on hand, amounts used, and the names of the workers involved (40, 229).

(f) Prevention of Spills and Accidents: Be prepared for accidents and spills (41).

Assure that at least 2 people are present at all times if a compound in use is highly toxic or of unknown toxicity (39).

Store breakable containers of these substances in chemically resistant trays; also work and mount apparatus above such trays or cover work and storage surfaces with removable, absorbent, plastic backed paper (40).

If a major spill occurs outside the hood, evacuate the area; assure that cleanup personnel wear suitable protective apparel and equipment (41).

(g) Waste: Thoroughly decontaminate or incinerate contaminated clothing or shoes (41). If possible, chemically decontaminate by chemical conversion (40).

Store contaminated waste in closed, suitably labeled, impervious containers (for liquids, in glass or plastic bottles half-filled with vermiculite) (40).

4. *Work with Chemicals of High Chronic Toxicity* (examples: dimethyl-

mercury and nickel carbonyl (48), benzo-a-pyrene (51), N-nitrosodiethylamine (54), other human carcinogens or substances with high carcinogenic potency in animals (38)).

Further supplemental rules to be followed, in addition to all these mentioned above, for work with substances of known high chronic toxicity (in quantities above a few milligrams to a few grams, depending on the substance) (47) (Procedure A of "Prudent Practices" pp. 47–50).

(a) Access: Conduct all transfers and work with these substances in a "controlled area": a restricted access hood, glove box, or portion of a lab, designated for use of highly toxic substances, for which all people with access are aware of the substances being used and necessary precautions (48).

(b) Approvals: Prepare a plan for use and disposal of these materials and obtain the approval of the laboratory supervisor (48).

(c) Non-contamination/Decontamination: Protect vacuum pumps against contamination by scrubbers or HEPA filters and vent them into the hood (49). Decontaminate vacuum pumps or other contaminated equipment, including glassware, in the hood before removing them from the controlled area (49, 50).

Decontaminate the controlled area before normal work is resumed there (50).

(d) Exiting: On leaving a controlled area, remove any protective apparel (placing it in an appropriate, labeled container) and thoroughly wash hands, forearms, face, and neck (49).

(e) Housekeeping: Use a wet mop or a vacuum cleaner equipped with a HEPA filter instead of dry sweeping if the toxic substance was a dry powder (50).

(f) Medical Surveillance: If using toxicologically significant quantities of such a substance on a regular basis (e.g., 3 times per week), consult a qualified physician concerning desirability of regular medical surveillance (50).

(g) Records: Keep accurate records of the amounts of these substances stored (229) and used, the dates of use, and names of users (48).

(h) Signs and Labels: Assure that the controlled area is conspicuously marked with warning and restricted access signs (49) and that all containers of these substances are appropriately labeled with identity and warning labels (48).

(i) Spills: Assure that contingency plans, equipment, and materials to minimize exposures of people and property in case of accident are available (233–4).

(j) Storage: Store containers of these chemicals only in a ventilated, limited access (48, 227, 229) area in appropriately labeled, unbreakable, chemically resistant, secondary containers (48, 229).

(k) Glove Boxes: For a negative pressure glove box, ventilation rate must be at least 2 volume changes/hour and pressure at least 0.5 inches of water (48). For a positive pressure glovebox, thoroughly check for leaks before each use (49). In either case, trap the exit gases or filter them through a HEPA filter and then release them into the hood (49).

(l) Waste: Use chemical decontamination whenever possible; ensure that containers of contaminated waste (including washings from contaminated flasks) are transferred from the controlled area in a secondary container under the supervision of authorized personnel (49, 50, 223).

5. *Animal Work with Chemicals of High Chronic Toxicity.*

(a) Access: For large scale studies, special facilities with restricted access are preferable (56).

(b) Administration of the Toxic Substance: When possible, administer the substance by injection or gavage instead of in the diet. If administration is in the diet, use a caging system under negative pressure or under laminar air flow directed toward HEPA filters (56).

(c) Aerosol Suppression: Devise procedures which minimize formation and dispersal of contaminated aerosols, including those from food, urine, and feces (e.g., use HEPA filtered vacuum equipment for cleaning, moisten contaminated bedding before removal from the cage, mix diets in closed containers in a hood) (55, 56).

(d) Personal Protection: When working in the animal room, wear plastic or rubber gloves, fully buttoned laboratory coat or jumpsuit and, if needed because of incomplete suppression of aerosols, other apparel and equipment (shoe and head coverings, respirator, etc.) (56).

(e) Waste Disposal: Dispose of contaminated animal tissues and excreta by incineration if the available incinerator can convert the contaminant to non-toxic products (238); otherwise, package the waste appropriately for burial in an EPA-approved site (239).

F. Safety Recommendations

The above recommendations from "Prudent Practices" do not include those which are directed primarily toward prevention of physical injury rather than toxic exposure. However, failure of precautions against injury will often have the secondary effect of causing toxic exposures. Therefore, we list below page references for recommendations concerning some of the major categories of safety hazards which also have implications for chemical hygiene:

1. Corrosive agents: (35–6)
2. Electrically powered laboratory apparatus: (179–92)
3. Fires, explosions: (26, 57–74, 162–4, 174–5, 219–20, 226–7)
4. Low temperature procedures: (26, 88)
5. Pressurized and vacuum operations (including use of compressed gas cylinders): (27, 75–101)

G. Material Safety Data Sheets

Material safety data sheets are presented in "Prudent Practices" for the chemicals listed below. (Asterisks denote that comprehensive material safety data sheets are provided.)

*Acetyl peroxide (105)
*Acrolein (106)
*Acrylonitrile (107)
Ammonia (anhydrous) (91)
*Aniline (109)
*Benzene (110)
*Benzo(a)pyrene (112)
*Bis(chloromethyl)ether (113)
Boron trichloride (91)
Boron trifluoride (92)
Bromine (114)
*Tert-butyl hydroperoxide (148)
*Carbon disulfide (116)
Carbon monoxide (92)
*Carbon tetrachloride (118)
*Chlorine (119)
Chlorine trifluoride (94)
*Chloroform (121)
*Chloromethane (93)
*Diethyl ether (122)
Diisopropyl fluorophosphate (41)
*Dimethylformamide (123)
*Dimethyl sulfate (125)
*Dioxane (126)
*Ethylene dibromide (128)
*Fluorine (95)
*Formaldehyde (130)
*Hydrazine and salts (132)
Hydrofluoric acid (43)
Hydrogen bromide (96)
Hydrogen chloride (96)
*Hydrogen cyanide (133)
*Hydrogen sulfide (135)
Mercury and compounds (52)
*Methanol (137)
*Morpholine (138)
*Nickel carbonyl (99)
*Nitrobenzene (139)
Nitrogen dioxide (100)
N-nitrosodiethylamine (54)
*Peracertic acid (141)
*Pehnol (142)
*Phosgene (143)
*Pyridine (144)
*Sodium azide (145)
*Sodium cyanide (147)
Sulfur dioxide (101)
*Trichloroethylene (149)
*Vinyl chloride (150)

Appendix B—References (Non-Mandatory)

The following references are provided to assist the employer in the development of a Chemical Hygiene Plan. The materials listed below are offered as non-mandatory guidance. References listed here do not imply specific endorsement of a book, opinion, technique, policy or a specific solution for a safety or health problem. Other references not listed here may better meet the needs of a specific laboratory.

(a) Materials for the development of the Chemical Hygiene Plan:

1. American Chemical Society, Safety in Academic Chemistry Laboratories, 4th edition, 1985.

2. Fawcett, H. H. and W. S. Wood, Safety and Accident Prevention in Chemical Operations, 2nd edition, Wiley-Interscience, New York, 1982.

3. Flury, Patricia A., Environmental Health and Safety in the Hospital Laboratory, Charles C. Thomas Publisher, Springfield IL, 1978.

4. Green, Michael E. and Turk, Amos, Safety in Working with Chemicals, Macmillan Publishing Co., NY, 1978.

5. Kaufman, James A., Laboratory Safety Guidelines, Dow Chemical Co., Box 1713, Midland, MI 48640, 1977.

6. National Institutes of Health, NIH Guidelines for the Laboratory

use of Chemical Carcinogens, NIH Pub. No. 81–2385, GPO, Washington, DC, 20402 1981.

7. National Research Council, Prudent Practices for Disposal of Chemicals from Laboratories, National Academy Press, Washington, DC, 1983.

8. National Research Council, Prudent Practices for Handling Hazardous Chemicals in Laboratories, National Academy Press, Washington, DC, 1981.

9. Renfrew, Malcolm, Ed., Safety in the Chemical Laboratory, Vol. IV, *J. Chem. Ed.*, American Chemical Society, Easton, PA, 1981.

10. Steere, Norman V., Ed., Safety in the Chemical Laboratory, *J. Chem. Ed.*, American Chemical Society, Easton, PA 18042, Vol. I., 1967, Vol. II, 1971, Vol. III, 1974.

11. Steere, Norman V., Handbook of Laboratory Safety, the Chemical Rubber Company, Cleveland, OH, 1971.

(b) Hazardous Substances Information:

1. American Conference of Governmental Industrial Hygienists, Threshold Limit Values for Chemical Substances and Physical Agents in the Workroom Environment with Intended Changes, P.O. Box 1937, Cincinnati, OH 45201 (latest edition).

2. Annual Report on Carcinogens, National Toxicology Program, U.S. Department of Health and Human Services, Public Health Service, U.S. Government Printing Office, Washington, DC (latest edition).

3. Best Company, Best Safety Directory, Vols. I and II, Oldwick, NJ, 1981.

4. Bretherick, L., Handbook of Reactive Chemical Hazards, 2nd edition, Butterworths, London, 1979.

5. Bretherick, L., Ed., Hazards in the Chemical Laboratory, 3rd edition, Royal Society of Chemistry, London, 1981.

6. Code of Federal Regulations, 29 CFR Part 1910 Subpart Z. U.S. Govt. Printing Office, Washington, DC 26402 (latest edition).

7. IARC Monographs on the Evaluation of the Carcinogenic Risk of Chemicals to Man. World Health Organization Publications Center, 49 Sheridan Avenue, Albany, New York 12210 (latest editions).

8. NIOSH/OSHA Pocket Guide to Chemical Hazards. NIOSH Pub. No. 78–210, U.S. Government Printing Office, Washington, DC, 1978.

9. Occupational Health Guidelines, NIOSH/OSHA NIOSH Pub. No. 81–123, U.S. Government Printing Office, Washington, DC, 1981.

10. Patty, F. A., Industrial Hygiene and Toxicology, John Wiley & Sons, Inc., New York, NY (Five Volumes).

11. Registry of Toxic Effects of Chemical Substances. U.S. Department of Health and Human Services, Public Health Service, Centers for Disease Control, National Institute for Occupational Safety and Health, Revised Annually, for sale from Superintendent of Documents, U.S. Govt. Printing Office, Washington, DC 20402.

12. The Merck Index: An Encyclopedia of Chemicals and Drugs. Merck and Company Inc. Rahway, NJ, 1976 (or latest edition).

13. Sax, N. I. Dangerous Properties of Industrial Materials, 5th edition, Van Nostrand Reinhold, NY, 1979.

14. Sittig, Marshall, Handbook of Toxic and Hazardous Chemicals, Noyes Publications, Park Ridge, NJ, 1981.

(c) Information on Ventilation:

1. American Conference of Governmental Industrial Hygienists Industrial Ventilation, 16th edition, Lansing, MI, 1980.

2. American National Standards Institute, Inc., American National Standards Fundamentals Governing the Design and Operation of Local Exhaust Systems ANSI Z 9.2—1979 American National Standards Institute, NY 1979.

3. Imad, A. P. and Watson, C. L. Ventilation Index: An Easy Way to Decide about Hazardous Liquids. *Professional Safety*, pp. 15–18, April 1980.

4. National Fire Protection Association:

- Fire Protection for Laboratories Using Chemicals NFPA—45, 1982.
- Safety Standard for Laboratories in Health Related Institutions, NFPA, 56c, 1980.
- Fire Protection Guide on Hazardous Materials, 7th edition, 1978, National Fire Protection Association, Batterymarch Park, Quincy, MA 02269.

5. Scientific Apparatus Makers Association (SAMA), Standard for Laboratory Fume Hoods, SAMA LF7–1980, 1101 16th Street, NW., Washington, DC 20036.

Appendix 3

PUBLIC LAW 99–499, SEPTEMBER 17, 1986 SUPERFUND AMENDMENTS AND REAUTHORIZATION (SARA) TITLE III, SUBPART A

SEC. 301. Establishment of State Commissions, Planning Districts, and Local Committees.

(a) *Establishment of State Emergency Response Commissions.*—Not later than six months after the date of the enactment of this title, the Governor of each State shall appoint a State emergency response commission. The Governor may designate as the State emergency response commission one or more existing emergency response organizations that are State-sponsored or appointed. The Governor shall, to the extent practicable, appoint persons to the State emergency response commission who have technical expertise in the emergency response field. . . .

(b) *Establishment of Emergency Planning Districts.*—Not later than nine months after the date of the enactment of this title, the State emergency response commission shall designate emergency planning districts in order to facilitate preparation and implementation of emergency plans. . . .

(c) *Establishment of Local Emergency Planning Committees.*—Not later than 30 days after designation of emergency planning districts or 10 months after the date of the enactment of this title, whichever is earlier, the State emergency response commission shall appoint members of a local emergency planning committee for each emergency planning district. Each committee shall include, at a minimum, representatives from each of the following groups or organizations: elected State and local officials; law enforcement, civil defense, firefighting, first aid, health, local environmental, hospital, and transportation personnel; broadcast and print media; community groups; and owners and operators of facilities subject to the requirements of this subtitle. . . .

SEC. 302. Substances and Facilities Covered and Notification.

(a) *Substances Covered.*—

(1) *In general.*—A substance is subject to the requirements of this subtitle if the substance is on the list published under paragraph (2).

(2) *List of extremely hazardous substances.*—Within 30 days after the date of the enactment of this title, the Administrator shall publish a list of extremely hazardous substances. The list shall be the same as the list of substances published in November 1985 by the Administrator in Appendix A of the "Chemical Emergency Preparedness Program Interim Guidance."

(3) *Thresholds.*—(A) At the time the list referred to in paragraph (2) is published the Administrator shall—

(i) publish an interim final regulation establishing a threshold planning quantity for each substance on the list, taking into account the criteria described in paragraph (4), and

(ii) initiate a rulemaking in order to publish final regulations establishing a threshold planning quantity for each substance on the list.

(B) The threshold planning quantities may, at the Administrator's discretion, be based on classes of chemicals or categories of facilities.

(C) If the Administrator fails to publish an interim final regulation establishing a threshold planning quantity for a substance within 30 days after the date of the enactment of this title, the threshold planning quantity for the substance shall be 2 pounds until such time as the Administrator publishes regulations establishing a threshold for the substance. . . .

(b) *Facilities Covered.*—(1) Except as provided in section 304, a facility is subject to the requirements of this subtitle if a substance on the list referred to in subsection (a) is present at the facility in an amount in excess of the threshold planning quantity established for such substance. . . .

(c) *Emergency Planning Notification.*—Not later than seven months after the date of the enactment of this title, the owner or operator of each facility subject to the requirements of this subtitle by reason of subsection (b)(1) shall notify the State emergency response commission for the State in which such facility is located that such facility is subject to the requirements of this subtitle. Thereafter, if a substance on the list of extremely hazardous substances referred to in subsection (a) first becomes present at such facility in excess of the threshold planning quantity established for such substance, or if there is a revision of such list and the facility has present a substance on the revised list in excess of the threshold planning quantity established for such substance, the owner or operator of the facility shall notify the State emergency response commission and the local emergency planning committee within 60 days after such acquisition or revision that such facility is subject to the requirements of this subtitle.

SEC. 303. Comprehensive Emergency Response Plans.

(a) *Plan Required.*—Each local emergency planning committee shall complete preparation of an emergency plan in accordance with this section not later than two years after the date of the enactment of this title. The committee shall review such plan once a year, or more frequently as changed circumstances in the community or at any facility may require.

(b) *Resources.*—Each local emergency planning committee shall evaluate the need for resources necessary to develop, implement, and exercise the emergency plan, and shall make recommendations with respect to additional resources that may be required and the means for providing such additional resources.

(c) *Plan Provisions.*—Each emergency plan shall include (but is not limited to) each of the following:

(1) Identification of facilities subject to the requirements of this subtitle that are within the emergency planning district, identification of routes likely to be used for the transportation of substances on the list of extremely hazardous substances referred to in section 302(a), and identification of additional facilities contributing or subjected to additional risk due to their proximity to facilities subject to the requirements of this subtitle, such as hospitals or natural gas facilities.

(2) Methods and procedures to be followed by facility owners and operators and local emergency and medical personnel to respond to any release of such substances.

(3) Designation of a community emergency coordinator and facility emergency coordinators, who shall make determinations necessary to implement the plan.

(4) Procedures providing reliable, effective, and timely notification by the facility emergency coordinators and the community emergency coordinator to persons designated in the emergency plan, and to the public, that a release has occurred (consistent with the emergency notification requirements of section 304).

(5) Methods for determining the occurrence of a release, and the area or population likely to be affected by such release.

(6) A description of emergency equipment and facilities in the community and at each facility in the community subject to the requirements of this subtitle, and an identification of the persons responsible for such equipment and facilities.

(7) Evacuation plans, including provisions for a precautionary evacuation and alternative traffic routes.

(8) Training programs, including schedules for training of local emergency response and medical personnel.

(9) Methods and schedules for exercising the emergency plan.

(d) *Providing of Information.*—For each facility subject to the requirements of this subtitle:

(1) Within 30 days after establishment of a local emergency planning committee for the emergency planning district in which such facility is located, or within 11 months after the date of the enactment of this title, whichever is earlier, the owner or operator of the facility shall notify the emergency planning committee (or the Governor if there is no committee) of a facility representative who will participate in the emergency planning process as a facility emergency coordinator.

(2) The owner or operator of the facility shall promptly inform the emergency planning committee of any relevant changes occurring at such facility as such changes occur or are expected to occur.

(3) Upon request from the emergency planning committee, the

owner or operator of the facility shall promptly provide information to such committee necessary for developing and implementing the emergency plan.

SEC. 304. Emergency Notification.

(a) *Types of Releases.*—

(1) *302(a) Substance which requires CERCLA notice.*—If a release of an extremely hazardous substance referred to in section 302(a) occurs from a facility at which a hazardous chemical is produced, used, or stored, and such release requires a notification under section 103(a) of the Comprehensive Environmental Response, Compensation, and Liability Act of 1980 (hereafter in this section referred to as "CERCLA") (42 U.S.C. 9601 et seq.), the owner or operator of the facility shall immediately provide notice as described in subsection (b).

(2) *Other 302(a) substance.*—If a release of an extremely hazardous substance referred to in section 302(a) occurs from a facility at which a hazardous chemical is produced, used, or stored, and such release is not subject to the notification requirements under section 103(a) of CERCLA, the owner or operator of the facility shall immediately provide notice as described in subsection (b), but only if the release—

(A) is not a federally permitted release as defined in section 101(10) of CERCLA,

(B) is in an amount in excess of a quantity which the Administrator has determined (by regulation) requires notice, and

(C) occurs in a manner which would require notification under section 103(a) of CERCLA.

Unless and until superseded by regulations establishing a quantity for an extremely hazardous substance described in this paragraph, a quantity of 1 pound shall be deemed that quantity the release of which requires notice as described in subsection (b).

(3) *Non-302(a) substance which requires CERCLA notice.*—If a release of a substance which is not on the list referred to in section 302(a) occurs at a facility at which a hazardous chemical is produced, used, or stored, and such release requires notification under section 103(a) of CERCLA, the owner or operator shall provide notice as follows:

(A) If the substance is one for which a reportable quantity has been established under section 102(a) of CERCLA, the owner or operator shall provide notice as described in subsection (b).

(B) If the substance is one for which a reportable quantity has not been established under section 102(a) of CERCLA—

(i) Until April 30, 1988, the owner or operator shall provide, for releases of one pound or more of the substance, the same notice to the community emergency coordinator for the local emergency planning committee, at the same time and in the same form, as notice is provided to the National Response Center under section 103(a) of CERCLA.

(ii) On and after April 30, 1988, the owner or operator shall provide, for releases of one pound or more of the substance, the notice as described in subsection (b).

(4) *Exempted releases.*—This section does not apply to any release which results in exposure to persons solely within the site or sites on which a facility is located.

(b) *Notification.*—

(1) *Recipients of notice.*—Notice required under subsection (a) shall be given immediately after the release by the owner or operator of a facility (by such means as telephone, radio, or in person) to the community emergency coordinator for the local emergency planning committees, if established pursuant to section 301(c), for any area likely to be affected by the release and to the State emergency planning commission of any State likely to be affected by the release. With respect to transportation of a substance subject to the requirements of this section, or storage incident to such transportation, the notice requirements of this section with respect to a release shall be satisfied by dialing 911 or, in the absence of a 911 emergency telephone number, calling the operator.

(2) *Contents.*—Notice required under subsection (a) shall include each of the following (to the extent known at the time of the notice and so long as no delay in responding to the emergency results):

(A) The chemical name or identity of any substance involved in the release.

(B) An indication of whether the substance is on the list referred to in section 302(a).

(C) An estimate of the quantity of any such substance that was released into the environment.

(D) The time and duration of the release.

(E) The medium or media into which the release occurred.

(F) Any known or anticipated acute or chronic health risks associated with the emergency and, where appropriate, advice regarding medical attention necessary for exposed individuals.

(G) Proper precautions to take as a result of the release, including evacuation (unless such information is readily available to the community emergency coordinator pursuant to the emergency plan).

(H) The name and telephone number of the person or persons to be contacted for further information.

(c) *Followup Emergency Notice.*—As soon as practicable after a release which requires notice under subsection (a), such owner or operator shall provide a written followup emergency notice (or notices, as more information becomes available) setting forth and updating the information required under subsection (b), and including additional information with respect to—

(1) actions taken to respond to and contain the release,

(2) any known or anticipated acute or chronic health risks associated with the release, and

(3) where appropriate, advice regarding medical attention necessary for exposed individuals.

(d) *Transportation Exemption Not Applicable.*—The exemption provided in section 327 (relating to transportation) does not apply to this section.

SEC. 305. Emergency Training and Review of Emergency Systems.

(a) *Emergency Training.*—

(1) *Programs.*—Officials of the United States Government carrying out existing Federal programs for emergency training are authorized to specifically provide training and education programs for Federal, State, and local personnel in hazard mitigation, emergency preparedness, fire prevention and control, disaster response, long-term disaster recovery, national security, technological and natural hazards, and emergency processes. Such programs shall provide special emphasis for such training and education with respect to hazardous chemicals.

(2) *State and local program support.*—There is authorized to be appropriated to the Federal Emergency Management Agency for each of the fiscal years 1987, 1988, 1989, and 1990, $5,000,000 for making grants to support programs of State and local governments, and to support university-sponsored programs, which are designed to improve emergency planning, preparedness, mitigation, response, and recovery capabilities. Such programs shall provide special emphasis with respect to emergencies associated with hazardous chemicals. Such grants may not exceed 80 percent of the cost of any such program. The remaining 20 percent of such costs shall be funded from non-Federal sources.

Bibliography

American Chemical Society. 1980. *Forum on Hazardous Waste Management at Academic Institutions.* Washington, D.C.: ACS.

———. 1985a. *Less is Better—Laboratory Chemical Management for Waste Reduction.* Washington, D.C.: ACS.

———. 1985b. *Safety in Academic Chemistry Laboratories.* Washington, D.C.: ACS.

Berardinelli, S. P., and R. Hall. 1985. Site specific whole glove chemical permeation. *Am. Ind. Hyg. Assoc. J.* 46(2): 60.

Beyler, R. E., and V. K. Meyers. 1982. What every chemist should know about teratogens. *J. Chem. Ed.* 59: 759.

Bretherick, L., ed. 1961. *Hazards in the Chemical Laboratory.* London: Royal Society of Chemistry.

Bretherick, L. 1979. *Handbook of Reactive Chemical Hazards.* 2d ed. London: Butterworths.

Butcher, S. S., D. D. W. Mayo, R. M. Pike, C. M. Foote, J. R. Hotham, and D. S. Page. 1985. Micro scale organic laboratory I: An approach to improving instruction laboratory air quality. *J. Chem. Ed.* 62: 142.

Clean Water Act. 1978. Priority Pollutants. Pub. L. 95–217, revision.

Code of Federal Regulations. 1986. 40 CFR part 261.

CPC base—The Chemical Protective Clothing Data Base. 1985. Cambridge, Mass.: Arthur D. Little, Inc.

Fawcett, H. H. 1984. *Hazardous and Toxic Materials: Safe Handling and Disposal.* New York: Wiley/Interscience.

Fawcett, H. H., and W. S. Wood. 1982. *Safety and Accident Prevention in Chemical Operations.* 2d ed. New York: Wiley/Interscience.

Federal Register. 1980. RCRA Appendix VIII. 45, no. 98, p. 33132 (May 19).

Federal Register. 1986a. Rules and Regulations. 51, no. 216 (November 7).

Federal Register. 1986b. Superfund List. 51, no. 221, p. 41582 (November 17).

Fishbein, L. 1979. *Potential Industrial Carcinogens and Mutagens.* New York: Elsevier.

Fuscaldo, A. A., B. J. Erlick, and B. Hindman. 1980. *Laboratory Safety—Theory and Practice.* New York: Academic Press.

Garland, C. 1986. Chemical contamination and decontamination of protective clothing. Paper presented at American Industrial Hygiene Conference, 19 May, Dallas.

Green, M. A., and A. Turk. 1978. *Safety in Working with Chemicals.* New York: Macmillan.

Heinrich, H. W. 1931. *Industrial Accident Prevention.* New York: McGraw Hill.

Kurnath, N. T., and M. C. Kurnath. 1981. Chemistry laboratory litigations. *J. Chem. Ed.* 58: A329.

Merck & Co., Inc. 1983. *The Merck Index.* 10th ed. Rahway, N.J.: Merck & Co.

National Fire Protection Association. 1978a. *Fire Protection for Laboratories Using Chemicals.* Boston: NFPA.

———. 1978b. *Hazardous Chemical Reactions.* NFPA Manual 591-M. Boston: NFPA.

———. 1984. *Flammable and Combustible Liquids Code Handbook.* Boston: NFPA.

National Research Council. 1981. *Prudent Practices for Handling Hazardous Chemicals in Laboratories.* Washington, D.C.: National Academy Press.

———. 1983. *Prudent Practices for the Disposal of Hazardous Chemicals from the Laboratory.* Washington, D.C.: National Academy Press.

National Safety Council. 1983. *Accident Investigation: A New Approach.* Chicago: NSC.

Occupational Health and Safety. 1984. October, 59–62.

Pipitone, D. A., ed. 1984. *Safe Storage of Laboratory Chemicals.* New York: Wiley/Interscience.

Pipitone, D. A., and D. D. Hedberg. 1982. Safe chemical storage: A pound of prevention is worth a ton of trouble. *J. Chem. Ed.* 59: A159.

Pitt, M. J. 1984. Please do not touch: Some thoughts on temporary labels in the laboratory. *J. Chem. Ed.* 61: A231.

Sax, N. I. 1984. *Dangerous Properties of Industrial Materials.* 6th ed. New York: Van Nostrand Reinhold.

Sittig, M., ed. 1981. *Handbook of Toxic and Hazardous Chemicals.* Cleveland: Chemical Rubber Publishing Co.

Steere, N. J., ed. 1971. *CRC Handbook of Laboratory Safety.* 2d ed. Boca Raton, Fla.: CRC Press.

Threshold Limit Values for Chemical Substances and Physical Changes in the Workroom Environment. Cincinnati: American Conference of Governmental and Industrial Hygienists. Issued annually.

U.S. Environmental Protection Agency. 1986. *Understanding the Small Quantity Generator Hazardous Waste Rules.* EPA/530-SW-86-019. Washington, D.C.

U.S. Department of Health, Education and Welfare. 1977a. *Carcinogens—Regulation and Control.* DHEW Publication no. 77-205. Washington, D.C.: Government Printing Office.

———. 1977b. *Carcinogens—Working with Carcinogens.* DHEW Publication no. 77-206. Washington, D.C.: Government Printing Office.

Walters, D. B., ed. 1980. *Safe Handling of Chemical Carcinogens, Mutagens, Teratogens and Highly Toxic Substances.* 2 vols. Ann Arbor, Mich.: Ann Arbor Science.

Young, J. A. 1983. Academic laboratory waste disposal. *J. Chem. Ed.* 60: 490.

Index

NOTES

NOTES